Thomas Antisell

The Manufacture of Photogenic or Hydro-Carbon Oils

From Coal and Other Bituminous Substances, Capable of Supplying Burning Fluids

Thomas Antisell

The Manufacture of Photogenic or Hydro-Carbon Oils
From Coal and Other Bituminous Substances, Capable of Supplying Burning Fluids

ISBN/EAN: 9783337062415

Printed in Europe, USA, Canada, Australia, Japan

Cover: Foto ©berggeist007 / pixelio.de

More available books at **www.hansebooks.com**

THE MANUFACTURE

OF

PHOTOGENIC OR HYDRO-CARBON

OILS,

From Coal and other Bituminous Substances,

CAPABLE OF

SUPPLYING BURNING FLUIDS.

BY

THOMAS ANTISELL, M.D.,

PROFESSOR OF CHEMISTRY IN THE MEDICAL DEPARTMENT OF GEORGETOWN
COLLEGE, D. C., ETC., ETC.

NEW YORK:
D. APPLETON AND COMPANY,
346 & 348 BROADWAY.
LONDON: 16 LITTLE BRITAIN.
1860.

ENTERED, according to Act of Congress, in the year 1859, by
D. APPLETON & COMPANY,
In the Clerk's Office of the District Court of the United States for the
Southern District of New York.

PREFACE.

The present little treatise is the first published monograph on the art of distilling oils from minerals containing Bitumen: like the art itself, it is necessarily imperfect in some particulars. The difficulty of obtaining detailed information on methods of manufacture abroad or at home, is not inconsiderable, when the history and progress of an art has to be newly described.

The position which the Author occupies in the U. S. Patent Office (having in charge the examination of a large class of patented applications, involving chemical processes), has enabled him to present to the public this record of the origin and condition of an infant art—the well-furnished Library of the Patent Office having furnished him the means to indicate the state of the manufacture abroad.

It is hoped it will be acceptably received by those occupied with, or interested in, this new branch of industry.

CONTENTS.

CHAPTER I.

INTRODUCTION—HISTORY OF THE ART, . . 7

CHAPTER II.

ON THE CHEMICAL COMPOSITION OF BITUMINOUS COAL, BITUMINOUS SCHIST, NATIVE BITUMENS, PEAT, AND ORGANIC SUBSTANCES YIELDING PHOTOGENIC OILS, ... 17

CHAPTER III.

ON THE GENERAL PRINCIPLES INVOLVED IN DESTRUCTIVE DISTILLATION: RESULTING PRODUCTS OBTAINED, . . . 39

CHAPTER IV.

ON THE PRODUCTS DERIVED FROM THE DISTILLATION OF BITUMINOUS COAL, ... 47

CHAPTER V.

ON THE PRODUCTS DERIVED FROM THE DISTILLATION OF SCHISTS AND NATURAL BITUMENS, ... 77

CHAPTER VI.

OF THE DISTILLATION OF PEAT AND WOOD, . . . 85

CHAPTER VII.

ON THE VARIOUS MODES OF APPLYING HEAT IN THE PROCESS OF
DISTILLING PHOTOGENIC OILS, ... 92

CHAPTER VIII.

GENERAL REMARKS ON THE COMMERCIAL MANUFACTURE, . . 113
SYNOPTICAL RESUME OF PATENTED IMPROVEMENTS HAVING REFERENCE TO THE DISTILLATION OF OILS FROM COALS, BITUMENS,
AND SCHISTS, ... 136
    I. AMERICAN PATENTS, ... 136
    II. EUROPEAN PATENTS, ... 141

CHAPTER I.

HISTORICAL INTRODUCTION.

THE new and extensive manufacture of oils from coal and other bituminous substances, is one of the latest applications of that valuable mineral to new and important uses; and though still in its rudest infancy, it promises to become one of the most enlarged and valuable applications to which coal has been subjected.

When the number of products derivable from the destructive distillation of coal at low temperatures is taken into account, the many and varied uses to which each and all of these are capable of being adapted, the cheapness of production, and the unlimited capability of supply, we are tempted to believe that this last effort to further utilize an already inconceivably useful mineral, is the happiest modern result of the application of Chemistry to the arts of life.

The discovery of the production of oils from coal, appears to date as far back as the time of Boyle, when the experiments of Dr. Clayton were made upon the inflammable nature of the distillates of coal. These were first communicated to the public by the Royal Society of London, many years after, in 1739 : "first, (says he,) there came over a flegm, *then a black oil*, and then likewise a spirit (gas) arose, which I could in no wise condense." This gas was such a matter of novelty and interest, that the appearance or nature of the oil was overlooked, and Clayton's experiments were wholly directed to the examination of the gas, and not of the fluid products.

Dr. Hales, in his Vegetable Statics, published in 1726, describing experiments conducted by him, mentions the production of a volatile oil, which he condensed in a vessel attached to the still.

Dr. Watson, Bishop of Llandaff, also describes the production of oils whenever coal is heated to redness in close vessels.—(*Philos. Trans.*, Vol. 41.)

Mr. Northern, of Leeds, (England,) in the year 1805, called public attention to the use of coal gas, and in the Monthly Magazine for April, 1805, writes : " I distilled in a retort 50 oz. of pit coal in a red heat, which gave 6 oz. of a liquid matter covered with oil more or less fluid as the heat was increased or diminished ; about 26 oz. of cinder remained in the retort ; the rest came over in the form of air as it was collected in the pneumatic apparatus. ❋ ❋ ❋ ❋ ❋ In the receiver I found a fluid of an acid taste, with a great quantity of oil, and at the bottom a substance resembling tar."

This passage contains the germ or basis of the manufacture of Volatile Oils from Coal, which was not further pursued until nearly the middle of the present century, when the demand for rapid solvents of Caoutchouc became so urgent, that new modes of obtaining benzule led to the distillation of tar for that purpose, and while separating benzule by fractional distillation, other valuable luminiferous agents were found to be present, or capable of being derived from the same crude fluid.

Before the application of coals to the manufacture of gas, the necessity which existed for the use of tar for various purposes by the English navy and mercantile marine, led to the carbonization of coals for the obtaining of tar therefrom; and though generally esteemed inferior for these purposes to wood tar, yet the scarcity of forests and consequent high price of the latter, led to a ready market for coal tar. The subsequent extended manufacture of gas, not only in England, but throughout the world, led to a large supply of tar, independent of its separate manufacture. The manufacture of coke in ovens led also to a smaller additional supply.

Laurent and Reichenbach had shown the results yielded by the distillation of tars, and Selligue in France applied this knowledge to the practical treatment of the bituminous schists of Autun, and, still later, to the *paper coal* and bituminous slate of the coal formation.

Selligue purified the oils so as to make burning fluids of them, and was the true introducer of that industry into France. Mansfield, at the close of 1847, obtained his patent for the separation and purification of volatile liquids from tar: the benzule, which he introduced into the English market, at once commanded a ready sale, from its known property of dissolving caoutchouc. Before the mastication of rubber was practised, its solution was the only known mode of separating its particles and enabling sheet rubber to be made; and as turpentine acted but slowly as a solvent, benzule was esteemed a valuable addition to the arts. Mansfield had pointed out its property of rendering air or gases luminous when saturated with its vapor, and naphthalized gas, as it was termed, became an article in domestic use. The other fluids did not make their way into the market as burning fluids, whether owing to their small production or not is difficult to say. This was the position of matters in 1848 and up to 1850.

About this time, James Young obtained a Scotch patent, and subsequently an English one, for the obtaining of Paraffine oils from coal: the practical results of his process were so promising that the treatment of coals for the obtaining the distilled oils has every year increased in importance.

This discovery of Young's was one of a class very common in the history of technological improvement: not one in which the improvement has been of that character to astonish by its novelty, or excite admiration by its magnitude; but, on the other hand, a small step in advance of previous applied knowledge, an advance so slight as hardly to have elicited any surprise at the time of publication.

Many years before 1848, it had been known that bituminous schists afforded on destructive distillation considerable quantities of oil, and efforts were not wanting both in France and England to turn these shales to practical advantage.

The experiments of the Hon. Robert Boyle upon coal, by which he obtained a gas, were the first efforts to separate an illuminating agent from that mineral; his attempts were not repeated, and his discoveries lay without any practical result for exactly 100 years, when Mr. Murdock, of Cornwall, England, lighted up his house at Redruth with illuminating gas.

On the continent of Europe, the high price of animal oils and fats, and the insufficient supply of vegetable fat oils, directed attention to the distillation of asphalt, bitumens, and bituminous schists, so as to obtain oils for illumination therefrom.

The manufacture of bituminous oils, so extensively carried out in Germany and France for the last 15 years, is of comparatively recent growth in Great Britain and this country, where the pursuit of whale fishing supplied the market with abundance of lamp oils.

It was not, therefore, by tentative essays upon coal or its crude tar alone, that the production of volatile oils was wholly perfected. From the time when Lavoisier had opened up the new and exact mode of examining material substances, the bitumens of Europe had engaged the attention of chemists. Theodore de Saussure distilled the asphaltic limestone of Travers, Neufchâtel, (Switzerland,) in 1819 and 1820; he obtained an oil therefrom, and found it identical with that from the petroleum of Amiano. Reichenbach, the proprietor of the chemical works of Tuhirico, Moravia, while examining the results of the dry distillation of beech wood, in 1829 and 1830, discovered paraffin; he derived it from the tar of the wood. It was found in a few years that paraffin also existed in the tarry matters distilled from other species of wood, and also in the tars arising from the distillation of bitumens, and ultimately in coal. In 1830–'31, Reichenbach discovered naphthalin; in 1831–'32, he described kreosote, piccamar, and pittakal, all of which were derived from the tar obtained by the dry distillation of woody matters.

To no one are we so much indebted for opening up true views of the results of close distillation of organic (vegetable) substances, as to Reichenbach. His name will ever be coupled with the early history of the production of oils from bituminous matters; and it must be acknowledged that for many years all our information on this subject was derived from his researches. In 1833 and 1834, he turned his attention to the distillation of coal in close vessels in contact with water, but without any practical results ; from 220 lbs. of coal, he only obtained little more than 9 ounces of volatile oil, or about $\frac{3}{10}$ of 1 per cent. In 1833, Dr. Bley distilled brown coal, and obtained a small quantity of volatile oil, besides some ammoniacal products.

The difficulty in adjusting the due degree of heat, no doubt led to the discouraging results (viewed in a practical light) of the distillation of coals and bitumens ; and the extensive use of these materials in the production of gas, drew away the attention of the chemist and the manufacturer from the problem of obtaining liquid products instead of permanent gases.

Still, however, various attempts were made to improve the apparatus for distilling ; and the retorts of Hompesch, and Beslay, and Rouen, of Gengembre, and others, show that from 1841, correct views as to the means for distilling, for separating the products, and for adjusting the temperature, had commenced to be entertained, although these views were not carried out by treatment always appropriate or successful.

In September, 1812, Mr. Lewitte breveted an apparatus for extracting tar from coal, the object being the application of the tar to varnishes, and modes of protecting surfaces : two circular furnaces, placed at each end of the apparatus, with fan or blasts to activate the fires ; the condensing apparatus between the furnaces, consists of a series of narrow passages in brick work, with a reservoir placed in the centre and front of the apparatus to receive the bitumen. The mode of operation was, to place $3\frac{1}{2}$ tons of coal in one furnace, and the communication with the other furnace shut off by a register, the coals being kindled by lighting some kindling placed below the coal on the hearth, and opening the blasts, in two hours the combustion becomes active, and the tar commences to distil over. Combustion lasts 24 hours, and when over, the register of that furnace is shut, and the opposite one lighted, and thus the condensers may be kept in constant use by alternate fires. Coal afforded 10 per cent. of tar by this mode of distillation; the residual coke was valuable for forges and iron furnace operations.

In 1824, Prosper and Charles Chervau breveted a process for extracting by distillation the bitumen which the rocks in the department of Saone and Loire contain abundantly. The mode of treatment was to place the rock, broken into small pieces, into cylindrical cast-iron retorts, 5 feet 2 inches long, 20 inches broad, and about $1\frac{1}{4}$ inch thick; this cylinder is closed at one end by the luted cap or lid in the usual way, and at the other, the lid has an opening in its centre for the admission of the eduction tube, which passes into a receiver, or wolfe's bottle; in connection with this, nine other receivers are attached. The vessels are of stone ware, and so arranged by connecting pipes, that when the first receiver is half full, a pipe leads off the tar into the second, and so on until the last receiver is filled, when it is drawn off by a faucet. The retort is placed on a furnace, whose wall supports it at either end, leaving the centre of the retort free for the flame to play on. The receiving vessels may be emptied by a syphon when the distillation is finished.

The bitumen obtained had all the character of naphtha, and the manufacturers recommend it as well suited for giving light in alcohol lamps; they also state, that by operating on the rock of Saone and Loire they have extracted volatile oils in the proportion of 40 parts of oil to 100 parts of rock.

Many impediments presented themselves in the practical manufacture of products from the dry distillation of wood and coal. Reichenbach had shown the mode of preparation of oils from vegetable matters, (fresh,) from tarry matters, and finally from the carbonizing of pit coal, but the products were always trifling, and therefore even the establishment of moderate factories was neither profitable nor inviting. "So remained paraffin until this hour, a beautiful item in the collection of chemical preparations, but it has never escaped from the rooms of the scientific man."* Thus wrote Reichenbach of it in 1854. Only since the year 1850 has the manufacture of paraffin from pit coal, turf, and bituminous shales, succeeded as an art. The first manufacture was that of James Young, in Manchester, by whose process, from 100 parts of Cannel coal, 40 per cent. of oil and 10 per cent. of paraffin could be obtained.

In thus showing that the practical manufacture of oils from coal is due to James Young, it may not be amiss to call attention to what it was which he produced from coals by distillation. He claimed the production of paraffin oils —not the production of naphtha or benzule, nor naphthalin, but paraffin and its congeners: this involves the slower distillation of coals at a lower temperature than had been hitherto effected, and this novelty in practice was followed by the novel result of a copious production of isomeric liquid hydrocarbons; so that really two great results were first demonstrated practically by the operation of Young's process, namely—1st. That coal was a material from which liquids could be manufactured economically, as tar,

* Reichenbach Journ. f. pract. Chem. LXIII., p. 63.

bitumens, and schists, had been hitherto employed ; and 2d. That the liquids so formed were paraffin—containing compounds.

An impression has taken hold of the American manufacturing public, that the patent of James Young has no force, as it was not a new invention at the date of the patent ; and from the unfavorable effect of that patent upon the actual manufacture of coal oils in this country an ill-feeling has been produced against it. That the owners of this patent have not acted wisely by withholding sales and licenses under it until very lately, is to be regretted ; but that it was a *bona fide* improvement in an art at the time when it was patented, and that, therefore, the patent was rightly issued in this country, there can be no shadow of doubt in the mind of any one who carefully traces the steps of the discovery of the production of photogenic oils from different materials.

Chemists at the present day look upon the fluid bitumen from native sources, and the bitumen existing ready formed in coal, as substances which, if not identical, are very closely allied, so closely that both, when treated alike, yield products closely resembling each other. But chemists and naturalists did not always hold this opinion, and it was by no means a certainly ascertained fact, that the substances treated alike would yield like results ; in fact, the term bitumen applied to native plastic liquids and to the material in coal so named, conveyed not the same idea, but merely a remote resemblance ; it was a resemblance of physical rather than of chemical properties, and hence the fact propounded by Reichenbach, and practically demonstrated by Young, that bituminous coal on distillation yielded paraffin oils, was a considerable step in advance both in chemistry and manufacture.

In Germany there originated, in 1855, paraffin works at Beuel near Bonn, Ludwigshafen, and Toplitz. A few years has sufficed to introduce this waxy matter into many uses about us. This is shown in the more numerous modes, and the lower prices at which it is now obtained.

The manufacture thus established in Germany, was also founded in France and Austria by Selligue, and the success resulting called the attention of England and this country to it as a branch of manufacture.

The first manufacture in this country was the attempt of Solomon Gesner on the bituminous shales of Dorchester, New Brunswick. Extensive manufactories are now established at Brooklyn, New York, Pittsburg, Baltimore, and at several places along the Ohio valley and river. Yet the demand is so much in advance of the supply, that not only is the quantity produced insufficient, but the oil is sent into the market in such a crude and impure state, that much of the tar is retained, and the oil smokes and gives off unpleasant odors in the apartments.

This manufacture once established must always progress : the oils are valuable as solvents and as lubricators, as well as for photogenic purposes; in the latter use, they give *ceteris paribus* a whiter and a more brilliant light than any fixed or fat oil, and are produced at much less cost than oil can be had for. Hence, while they narrow the demand for fish and lard oils, which they supersede, and thus prevent the cost of such oils rising to any unusual price, they are themselves controlled by the price of oil; and it only requires sufficient attention to be bestowed upon its purification so as to free it from creosote impurities to render it one of the most pleasing and brilliant, as well as the most economic source of light in those situations where gas is not desirable or attainable

CHAPTER II.

OF THE NATURE OF COALS, CARBONACEOUS SCHISTS, NATIVE BITUMENS, AND ORGANIC SUBSTANCES YIELDING MINERAL OILS.

COAL is defined by Redfern to be a compressed and chemically altered vegetal matter, associated with more or less earthy substances, and capable of being used as fuel.

This restricts the origin of coal to vegetable substances, and perhaps with propriety, for we do not know of animal substances by their decomposition producing a substance having all the properties of coal.

Dr. Aitken, of Glasgow, and other microscopists, have carefully examined coal under the microscope, and in every case found traces of vegetable cells or structure, showing its plant-origin. Even in the most altered coals this could be ascertained; hence, in the hands of a skilful microscopical chemist, this test may be applied to determine with certainty whether the substance is a coal or a bitumen.

Native bitumens, asphalt, and petroleum, may have been formed also solely from vegetable matter undergoing decomposition under peculiar circumstances. A few geologists and chemists are willing, however, to admit that bitumens may be of animal origin, and in a few instances may have been produced by the slow subterranean alteration of fish remains deposited during former geologic periods. It is difficult to speak with certainty of the exact origin of bitumens—but in one respect they differ from coal. In no case can an organic tissue or structure be demonstrated when they are examined under the microscope.

Viewed in this light, the mineral found at the Albert mine, New Brunswick, should be classed as a bitumen, since Dr. J. Leidy was unable to detect any traces of structure in its mass : its difficulty of fusion is no argument against its being a bitumen, since many of the bitumens of France are not fusible. The chemists and mineralogists of this country have, however, generally agreed to class it with the Boghead coal of Scotland, as a variety of cannel coal.

The various changes or steps of the decomposition by which vegetable matter or wood is ultimately converted into coal, are not fully known. That time plays a considerable part in it, is evident from the difference between the true coals and the lignites, and even between lignites of different ages: the vascular and cellular characters of the wood being more evident in the lignites than in those coals in which the time for producing the change was prolonged. It is well known that carbonic acid gas escapes abundantly from faults and fissures in the beds of brown coal, which may be the source of the acidulous springs found in those neighborhoods. This loss of carbonic acid appears to accompany the conversion of wood into lignite, and the following formula, according to Gregory, would explain this occurrence :

NATURE OF COAL. 19

From 3 equivalents of wood C_{36} H_{22} O_{22} take
3 equivalents carbonic acid C_3 O_6
and 1 equiv. of hydrogen H_2
——————
There will be left of brown coal, C_{33} H_{21} O_{16}

The change here is the great loss of oxygen, which consequently relatively increases the proportion of hydrogen and carbon in lignite above ordinary wood.—The oxygen is removed as carbonic acid.

The further change into mineral coal appears to be accompanied with additional loss of carbonic acid—some watery vapor and a quantity of hydrogen which comes away united with carbon as carburetted hydrogen : this is known as the fire-damp of miners. Splint coal and cannel coal both have the composition C_{24} H_{13} O, in which the loss of both oxygen and hydrogen is evident, especially the former.

If we take the sum of these substances escaping, viz. :—

3 equivalents carburetted hydrogen, C_3 H_6
3 equivalents water, H_3 O_3
9 equivalents carbonic acid, C_9 O_{18}
——————
If this be deducted C_{12} H_9 O_{21}
from the formula of wood, C_{36} H_{22} O_{22}
——————
there would remain the formula of Cannel coal $=$ C_{24} H_{13} O

Caking coal may be represented as Cannel coal $=$ C_{24} H_{13} O, minus olefiant gas C_4 H_4, and has the formula C_{20} H_9 O.

The ultimate constitution of coal being pointed out, an interesting question presents itself—What is the state or condition in which the elements are contained in the coal ? The belief of many is, that one portion of the carbon is in a free or uncombined state, while the other and smaller portion is united with the hydrogen and oxygen to form what is known as the bituminous portion; and that the first effect of heat is to simply separate the combined from the uncombined carbon, and that the formation of anthracite is explained in this way; this may be the case with some lignites, but when the homogeneous mass which true coal presents when heated—its semifusibility—is considered, it would rather appear that the carbon is altogether combined into one proximate substance, and that the effect of heat is not to separate, but to decompose.

Do any of the substances found in the receiver after the distillation of coal originally exist in it? When coal is digested with ether but a small portion dissolves, which gives a brown color to the liquid, and which, when dry, has some of the characters of bitumen; but neither naphtha nor petrolene, which exist naturally in bitumens, can be separated by any solvent from coal, and it is therefore not likely that these substances exist in the fresh coal: we may suppose the latter to contain a series of carbides of hydrogen not yet separable by any of the solvents applied. When heated, these resolve themselves into tar at first and afterwards into volatile oils; this unknown substance is called bitumen—not that it has been proved to be identical with native bitumens, for it has not, but because that by distillation it affords some of the products which bitumen yields when similarly treated. Paraffin probably exists ready formed in some coals.

It is the loose application of the term *Bitumen* which has obscured the history of the improvement in the art which we are treating of. It is of late years only that chemists have come to look upon native bitumen, and that substance found in coals, as belonging to the same species or group: but twenty years ago this resemblance was not so apparent, and at that time, to attempt to apply coals to produce the same substances as bitumen was known to do, was not thought of; and although, as we have shown in the historical chapter, that these oils were actually capable of being produced by distillation of coal; yet that such was a necessary, constant, and reliable result was not apprehended. Hence the attempt of Mr. Young to produce these oils on the large scale from the distillation of coal was not put in operation, because not believed capable of resulting in a successful manufacture. It had been already known that bitumens and bituminous schists would yield these photogenic oils in abundance; it was also shown that birch tar, beech tar, and even coal tar, would also yield them; but the manufacture of such from coal was not thought of before Mr. Young's patent, because it was not known or believed that bituminous coal could yield them on a large scale. This Mr. Young accomplished, and thereby is entitled to the merit of producing oils from coal by distillation so as to establish a branch of industry : and this discovery of the value of coal is accorded to him by Reichenbach himself.

The proportion of bitumen present in coal varies with the amount of change which the vegetation has undergone since its deposition: actions of internal terrestrial heat, accompanied by moisture under great pressure, exerted upon it, tends to expel its bitumen, and reduce the coal to the condition of anthracite. All coal would be bituminous, were it not for these changes produced by geological alterations of the sedimentary strata in which the deposit took place.

The bitumen varies in amount from 10 to 63 per cent., the coal being termed fat in the latter, and dry in the former case.

Bituminous coal is softer and less lustrous than anthracite, of a black or brown-black color, and with a specific gravity ranging from 1.14 to 1.5. When the bitumen is in abundance, it is often difficult to say whether the substance is a coal or a bitumen.

The bitumen of coal resembles that afforded by nature, as asphalt and mineral tar, in its sensible qualities and general appearance ; it does not, however, contain the same proximate principles ; it does not yield petroline, nor does it by dry distillation yield the same fluids : they belong, however, to the same natural group or series, and tend to strengthen the opinion generally held that bitumen, petroleum, and asphalt arise from the decomposition of fossil vegetation. In some cases, as before stated, it may be true that the slow decomposition of animal matter may produce a similar substance ; the fossiliferous shales in many bituminous districts containing abundant exuviæ of molluscs and fishes, the decomposition of which in great abundance, under the peculiar circumstances in which they are placed, may produce a hydro-carbon bitumen similar to the mineral tar.

The natural bitumens always contain some volatile oil ready formed, and their varieties depend on the greater or less proportion of this volatile oil present in them.

Interspersed through the masses of coal are found small quantities of a great variety of bodies—carbo-hydrogens—resembling the oils and stearopten of plants closely in properties and combination. Thus ozocerite or fossil wax is found in cavities in the rock lying on the coal ; it is brown, of a foliated structure, fuses at 143°. Paraffine is also found associated with coal. Both of them have the same composition as olefiant gas.—(*Kane.*)

Mineral coal is generally, for purposes of scientific technical description, divided into three classes:
1. Anthracite or Glance coal
2. Lignite or Brown coal.
3. Black or Bituminous coal.

For distillation, the latter class is almost universally employed; the lignite having only a limited area of distribution and employment, and the anthracite not yielding any volatile liquids. A small quantity of a bituminous mineral, known as Boghead coal, has been employed in Great Britain, but its nature is not sufficiently well determined to regard it as a true coal, though usually classed with it.

The varieties of black coal are very numerous, but the great majority may be included under four great divisions, viz.:—

1. *Caking Coal:* melting when heated, and agglutinating in masses.

2. *Splint Coal:* possessing a splintery fracture.

3. *Cherry Coal:* burns without caking in any degree.

4. *Cannel Coal:* hard, compact, bituminous, inflames at a candle.

Generally speaking, the value of these varieties is in the order set down, the last variety yielding the largest amount of volatile matters; then the cherry coal, while the caking coal yields the least supply.

Cannel coal may occur in veins in different positions—that is, the cannel coal may occupy an inferior vein, while the ordinary pit coal may lie in the vein above; but when the two species of coal occur together, the cannel coal is much more frequently found lying above, and in contact with, the pit coal.

Breckenridge Cannel coal. This coal, in Breckenridge Co., Kentucky, is well exposed; the coal measures, reaching an elevation of 500 feet above mean high water of the Ohio, at Cloverport. Of the three coal seams found in that region, the two lower are common bituminous coal; the uppermost bed lies 300 feet above the level of the river, and is capped by a thick overlying mass of sandstone and shale. This bed is three feet in thickness, with a bituminous shale of some ten feet in addition, on which it immediately rests. The following are the characters of this coal, as given by Professor B. Silliman, Jr., in a paper read before the American Association for the Advancement of Science, May, 1854, from which this account is condensed:—

"1. Specific gravity, 1.14 to 1.16. Common bituminous coal varies from 1.27 to 1.35, and anthracite, from 1.50 to 1.85. The only coal lighter than this, so far as is known, is the so-called Albert coal, of New Brunswick, whose density is 1.13. The cause of this low density will be sought chiefly in the very large amount of volatile matter.

"2. Its tenacity and elasticity. Coals are usually brittle and inelastic: this is tough, and resists powerful and repeated blows, and rebounds the hammer like wood. The splints of this coal may be sensibly bent by pressure, and regain their original form again. A fissure in it may be sprung open by a wedge, and will close again on withdrawing it. The writer has never seen any other coal with this peculiarity.

"3. Its electrical power. This coal becomes powerfully excited by friction with resinous electricity. This peculiarity may be demonstrated very easily, and has never before been noticed in any other coal, so far as the writer has been able to learn, except in the 'Albert coal' of New Brunswick, before named. It is not easy to understand why other very highly bituminous coals should not have this property, but such is the fact with a large number that have been tried.

"4. Chemical constitution. This has been determined in the usual way by destructive distillation, with the following results, viz.: in 100 parts we have—

	(1)	(2)
Volatile at redness,	60.27	63.520
Fixed carbon,	31.05	27.160
Ash,	8.66	8.470
Hygroscopic moisture,		.777
Sulphur,	trace	trace
	99.98	99.927
Coke,	39.71	36.68

"A comparison of these analyses with those of other highly bituminous coals, will show that there are very few examples recorded of so high an amount of volatile matter. For example, we find, among American coals, that the 'Albert coal,' of New Brunswick, yields 61.74; the coal of Chippenville, Pa., 49.80; that of Kanawha, 41.85; of Pittsburg, 32.95, while the mean of the fat caking coals of Liverpool is 37.60 per cent. The Lowmoor Scotch Cannel, and the Boghead, also a Scotch coal, are the only ones giving a higher proportion of volatile matter. In fact, the ordinary proportions of volatile and fixed ingredients in bituminous coals are completely reversed in the Breckenridge Cannel."

By comparing this description of Cannel coal with the Boghead, the resemblance will be perceived, the Boghead mineral having perhaps less density and more volatile matter contained in it.

Among the varieties of Cannel coal mentioned here, should be the Boghead coal of Scotland. As the manufacture of Pyrogenous oils has introduced this mineral into extensive use as a raw material, an outline of its properties are here subjoined.

Boghead coal, or Torbane Hill mineral, is found in that group of marine deposits defined by geologists as the carboniferous limestone of the southern outcrop of the Firth of Forth coal field, and is worked on a large scale at Bathgate, near Edinburgh.

It is hard, brittle, of an earthy-black color, and breaks with a dull, even fracture. It burns with a bright, voluminous flame, and gives off much smoke. Its specific gravity is 1.155, being the lightest variety of European bituminous coal known; it has the following constitution, as determined by Drs. Penny (1) and Fyfe (2):—

	(1)	(2)
Volatile matters,	71.3	69.
Carbon in Coke,	11.3	9.25
Ash,	16.8	21.75
Moisture,	6	
	100.	100.

The ash, according to Dr. Fyfe, consists of 71 per cent. of silica, the rest being lime, magnesia, alumina, and a minute quantity of iron, in union with sulphur. The percentage of sulphur amounted to 0.13, equivalent to nearly 3 lbs. per ton of mineral. The larger amount of volatile matter contained in Boghead coal than is found in Cannel coals, is shown by the following comparative analyses, made by Dr. Penny:

	Specific Gravity.	Volatile Matters.	Carbon in Coke.	Ash.	Sulphur, per cent.
Boghead,	1.155	71.9	11.3	16.8	.3
Lismahago,	1.240	52.6	41.	6.4	.74
Scotch Cannel (average).	1.330	50.6	41.9	7.5	1.26
Breadisholm,	1.319	40.5	51.5	8.	.3
Derbyshire Cannel,		47.	48.4	4.6	

When distilled, the Boghead coal yields several products, among which the absence of benzine is remarkable, it being present in but small quantity in the first distillation.

C. G. Williams found that the naphtha arising from the distillation of Boghead coal has a very low density, being only .750 at 60° F., although the boiling point, previous to rectification, was 290° F.; by repeated fractional distillations and purifications by nitric acid and alkaline solutions, a colorless and mobile fluid, having the odor of hawthorn blossoms, was obtained, with a density of .725 at 60°; this was butyle; he also separated propyle, amyle, and caproyle. The following are the chief properties of these liquids, as obtained from Torbane mineral by Williams:—

		Boiling Point.	Specific Gravity.	Vapor, density found.
Propyle,	$C_{12} H_{14}$	68. C.	.6745	2.96
Butyle,	$C_{16} H_{18}$	119. C.	.6945	3.88
Amyle,	$C_{20} H_{22}$	159. C.	.7365	4.93
Caproyle,	$C_{24} H_{26}$	202. C.	.7568	5.83

Paraffine, picoline, and phenole, have been obtained from the distillate of this mineral by Dr. Genther, whose results rather support the view that it is not a true coal: that the bituminous matter in it is not identical with that of ordinary black coal is evident from the different products obtained by distillation; in chemical constitution, it is more properly related to the petroleums and true bitumens. C. G. Williams has detected hexylene, $C_{12} H_{12}$, and heptylene, $C_{14} H_{14}$, two volatile hydro-carbon liquids.

It is not intended in this work to enter into a description of the variety and extent of the coal beds of the United States; the reader is referred to Taylor's Statistics of Coal, and to the works of W. R. Johnson.

Dr. Newberry looks upon cannel coal and bituminous shale as but variations of one substance, the coal being changed into a shale by the addition of earthy matter. That cannel coal has been formed by the constant submergence of vegetable matter under water, which, with the great pressure accompanying, gives it its homogeneous and laminated structure. The process of bituminization consists in the escape of some carbonic acid, of hydrogen as water, and the union of carbon and hydrogen to form the various hydro-carbons. Water preserves the coal from too much oxidation, and hence cannel coal formed under water contains more bitumen than other coals. The presence of fish-remains in cannel coal, according to Dr. Newberry, shows its aquatic deposition, and also that it must contain a certain amount of animal matter. Splint and cannel coals are formed somewhat differently from brown coal or lignite, the change taking place without access of air—consisting in the removal of 3 eq. of carburetted hydrogen, 3 eq. of water, and 9 eq. of carbonic acid, thus:

$$\begin{array}{ll} \text{Woody fibre,} & C_{36} H_{22} O_{22} \\ \text{3 eq. carburetted hydrogen,} & C_3 H_6 \\ \text{3 eq. water,} & H_3 O_3 \\ \text{9 eq. carbonic acid,} & C_9 \quad O_{18} \quad C_{12} H_9 O_{21} \\ \hline \text{Splint and cannel coal,} & C_{24} H_{13} O \end{array}$$

The following are the localities of cannel coal, as given by R. C. Taylor:

Virginia—Near Charleston, Kanawha Co., and on Kanawha River and its tributaries, ... ——
    Brandt's Mines, Potomac Valley, ... 5 feet
Pennsylvania—Near Greensburg, Beaver Co., . . . 8 feet
    Six miles west of Greensburg, ... ——
    Near Ebensburg, Cambria Co., ... ——
Kentucky—Breckenridge Co., Cloverport, ... 3 feet
Indiana—Cannelton, ... 3–5 feet
Missouri—St. Louis, 8 miles from, ... ——
    Callaway Co., ... 24–46 feet

The figures indicate the thickness of the vein.

The Breckenridge cannel coal contains, in 100 parts, according to the analyses of Dr. R. Peter (1), and B. Silliman, Jr. (2, 3):

	(1)	(2)	(3)
Moisture,	1.30	....	.777
Volatile matters,	54.40	60.27	63.525
Carbon,	32.	31.05	27.160
Ash,	12.30	8.66	8.470

LIGNITE or BROWN COAL is more recent in its formation than the upper secondary or true carboniferous deposits; it is sometimes not easily distinguishable from common bituminous coal. Usually of a brown-black color, bright coal lustre, with something of the texture of wood remaining—often the form and fibre of the original tree is retained, when it is called lignite : this latter burns with an empyreumatic odor. Brown coal occurs in beds usually of small extent, and is seldom so pure from pyrites as the more ancient bituminous coal.—(*Dana.*) The depth of the brown color depends upon the depth of the bed of coal itself; it rarely has a conchoidal fracture well defined ; it is usually inferior as a fuel, containing a large percentage of earthy matters. The coal contains usually a large amount of moisture, Varrentrop having found as much as 48 per cent. in a variety from Helmstadt; the average moisture is about 30 per cent., and when air-dried in summer, sinks to 20 per cent.

When lignite is treated with Caustic Potassa, it almost completely dissolves, yielding a dark brown liquor, which gives the same reactions that the presence of *Ulmine* in solution does.

As this is the chief ingredient of peat, it shows the close relation which exists between lignites and peat, closer than exists between lignites and bituminous coals. Again, the coke which remains after the distillation of lignites retains the form and structure of the sample operated on, as wood does, and which never occurs with other varieties of coal. As lignites are of various ages of deposit, the older varieties approach in characters very closely to bituminous coal.

Regnault gives the density of the various lignites which he examined when in the dry state, as between 1.100 and 1.85, the denser specimens containing a large amount of earthy matters.

Lignite is found in many places in the United States, not, however, in such quantity as to justify its being worked for the purposes of distillation.

The following table of the composition of American bituminous coal, with the ratio of fixed to volatile matters, is extracted from the report upon American coals made to the Navy Department of the United States, by W. R. Johnson; printed First Session of the Twenty-eighth Congress:—

Locality.	Specific Gravity.	Moisture.	Volatile Matters.	Fixed Carbon.	Earthy Matters.	Ratio of Volatile to Fixed Matters.
Pittsburg,	1.252	1.397	36.603	54.926	7.074	2.014
Cannelton, Indiana,	1.273	2.597	33.992	58.487	4.974	1.719
MARYLAND. Maryland N.Y. & Md. Ms.Co.	1.431	.808	12.902	67.365	18.980	5.222
Frostburg (Neff),	1.3221	2.455	12.675	74.527	10.343	5.858
Cumberland,	1.3092	1.070	15.173	77.252	6.520	5.096
do.	1.3050		17.411	77.350	5.239	4.733
do.	1.4023	.808	15.237	74.761	9.109	4.906
PENNSYLVANIA. Dauphin and Susquehanna,	1.4481	.446	18.577	74.214	11.494	....
Blossburg,	1.3236	1.839	18.927	73.108	10.778	
Ralston,	1.3949	.670	18.807	71.532	13.961	5.181
Quin's Run, Clinton Co.,	1.3404	.181	18.676	78.443	7.750	3.980
Carthaus, Clearfield Co.,	1.2919	.770	21.500	72.643	5.087	3.378
Cambria Co.,	1.3617	1.105	20.255	69.590	9.050	3.485
Barr's Run,	1.362	1.785	19.782	67.958	10.475	3.435
Crouch & Snead,	1.451	1.795	28.959	59.976	14.280	2.499
VIRGINIA. Eastern Coal fields. Midlothian,	1.437	1.172	27.278	61.088	10.467	2.239
Creek Co.,	1.319	1.450	29.678	60.300	8.572	2.082
Clover Hill,	1.285	1.839	31.698	56.881	10.182	1.793
Chesterfield, Ms. Co.,	1.289	1.896	30.676	58.794	8.634	1.917
do. do.	1.294	2.455	29.796	58.012	14.737	1.780
Tippecanoe,	1.346	1.841	34.165	54.620	9.374	1.590
Midlothian, new shaft,	1.325	.070	33.490	56.400	9.440	1.654
do. screened,	1.283	1.785	34.497	54.063	9.655	1.567
do. Navy Yard,	1.390	1.014	28.786	56.112	14.188	1.958

The same experimenter has given the results of his examination of foreign bituminous coal, which may serve as a point of comparison:—

NATURE OF BITUMEN. 31

Locality.	Specific Gravity.	Moisture.	Volatile Matter.	Fixed Carbon.	Earthy Matters.	Ratio of Volatile to Fixed Matter.
Picton No. 1,	1.318	2.567	27.063	56.981	13.389	2.105
do. No. 2,	1.825	.751	25.955	60.735	12.509	2.508
Sydney,	1.338	3.125	23.810	67.570	5.495	2.838
Liverpool,	1.262	.892	39.587	54.899	4.622	1.513
Newcastle,	1.257	2.007	87.597	56.996	5.400	1.601
Scotch,	1.519	3.018	38.537	48.812	9.333	1.257

Bitumens are viscous matters, ordinarily brown or black, which melt with facility at the temperature of boiling water, or even below that—but sometimes at a more elevated point; solid bitumens are named *asphaltes*. Bituminous deposits are the more or less metamorphosed products of organic life of a former geological period, which have been forced upwards through the incumbent clays or later deposits, and are found as wells or springs of viscid fluid, which harden at the surface and edges into a solid asphalt. Bitumen is found as a fluid or viscid mass in various parts of Europe; in France, in the Basaltic Tufa of Auvergne, in the Tertiary sands at Galion near Pézénas, at Lobbsan, and at Bechelbronn (Lower Rhine), in the upper cretaceous deposits, at Orthez, and at Cauperme near Dax, at Seysell near the junction of the Rhone and the Isere in Switzerland; wherever the Alpine limestone (Calcaire Alpien) is met with. In England, elastic bitumen occurs in Derbyshire.

Asphalt, which is not generally found in Europe, occurs in a thick bed at Arlona in Albania, but is chiefly derived from Lake Asphaltites or the Dead Sea; is found at Coquitambo, near Cuenca, Peru; in the West India islands, Barbadoes and Trinidad—in the latter place forming a lake three miles in circumference.

The deposits of bitumen on the American continent are perhaps as numerous as on the eastern. In the

United States, an extensive development occurs in the southern part of California, extending 400 miles along the western side of the coast range, close to the shores of the Pacific Ocean. At Santa Barbara it is found as a brittle, dark asphalt; at Los Angeles, it oozes up near the town, and forms a small lake of liquid petroleum, which slowly hardens. At San Luis Obispo, veins of bitumen intersect the strata which are there exposed and elevated, and in summer they become soft, and overflow over the surface of the rock; in winter they are almost unimpressible by the finger-nail.

Liquid bitumen was found on False Washita, near Washington Mountains, Kansas, by Lieut. Johnston, U. S. A., exuding from a dark sandstone. In Texas, within 100 miles of Houston, between Liberty and Beaumont, is a small bituminous lake, resembling the pitch lake of Trinidad : in winter it is covered with acidulous water; in summer, petroleum oozes up. Around Burksville, Ky., are several petroleum springs. Mather, in his reconnoissance of Kentucky, points out several places where petroleum might be collected.

Near Pittsburg, on the Alleghany river, a stream of petroleum was found in digging or boring for salt, which yields at times 1,800 bbls. per day from one point.

There are few bitumens which do not harden or become more solid by exposure to the atmosphere; this may arise from two causes: 1st. By the evaporation of the petroline or fluid naphtha portion, which the majority of bitumens contain ; and 2d. By the oxidation of the bitumen by exposure. Solid bitumen or asphalt always contains oxygen as an essential constituent, and the fresh fluid-like petroleum freshly poured out, which contains no oxygen, gradually acquires it as it hardens into asphalt.

The solid bitumen of the Dead Sea, and that of the Tar Lake of Trinidad, as well as the viscid varieties of that island, as also of many other parts, as at Bechelbronn in France, *et cetera*, are apparently produced from the oxidation of petroleum and their composition, as exhibited in the following formulæ :

Naphtha or petroleum, $C_{20} H_{16}$ or $C_{40} H_{32}$
Asphalt or bitumen, $C_{40} H_{32} O_6$

which manifests this derivation in a striking manner.— (*Muspratt.*)

The specific gravity of solid bitumens (asphaltes) ranges from 1. to 1.10 ; and no two specimens of bitumen, solid or liquid, can be said to have an identical chemical constitution.

Some bitumens are totally insoluble in alcohol; others are in part soluble, but none are wholly soluble ; most of them are acted on by ether or spirit of turpentine, and leave a carbonaceous residue, or some other bituminous matter not attacked by these solvents, and whose point of fusion is different from that of the primitive bitumen. When submitted to distillation, they are in part separated into the liquids originally present, and in part are decomposed into oleaginous liquids.

According to the researches of Boussingault, the semi-liquid bitumens are mixtures of two definite principles, *asphaltine* solid and fixed, and *petroline* fluid and volatile. These two substances may be separated by exposing the bitumen to the temperature of boiling water in a close vessel.

Petroline is a pale-yellow oil, having a sharp taste, and odor of bitumen ; specific gravity, .891 at 70°. It stains paper, and burns with a smoky flame ; it boils at 536°, and the density of the vapor is 9.415. It yields, on analysis, carbon, 87.3, and hydrogen, 11.9 in 100 parts, and may be represented by the formula $C_{40} H_{32}$, which corresponds to a vapor density of 9.50 (4 volumes). Its resemblance to naphtha is very close. Alcohol dissolves only a small portion of petroline, but the whole is removed by keeping the bitumen for 48 hours at a temperature of 472°. The asphaltine remaining behind is black, very brilliant, and has a conchoidal fracture; at 540° it becomes soft and elastic, and melts before decomposition; it burns like resin, and yields on analysis:

Carbon, 74.2
Hydrogen, 9.9 } represented by the formula $C_{40} H_{32} O_4$
Oxygen, 15.1

Asphaltine would thus appear to be merely the result of the oxidation of petroline.

The Burmese naphtha, or Rangoon petroleum, elsewhere alluded to in this treatise, is perhaps one of the most perfectly altered bitumens that has been brought under the notice of chemists. When distilled at 212° F., it yields volatile hydro-carbons without any substance reacting on it; and the substances which are thus separated have boiling points widely apart, some of them being above 400° F. As this temperature was not reached in the mere distillation, it may be presumed that these hydro-carbon oils pre-existed ready formed in the naphtha, and when the most volatile of these, the benzine, rises, it carries over with it the vapors of the other hydro-carbons; the vapors of all of this class readily diffusing with each other. The liquid having the lowest specific gravity, and which comes over first by distillation, is an analogue of benzine, and has been termed *Sherwoodole;* it has analogous properties in dissolving grease, &c. It may be eliminated, in the same manner as benzole, by means of sulphuric acid.

The steam distillate at 212° amounts to one-fourth of the whole crude liquid. The residue is treated with concentrated sulphuric acid, which purifies it from foreign matters (which De la Rue and Müller have shown to belong to the colophene series). This matter thrown down by the acid is of a black color, and seems to be in every respect identical with true asphaltum. The purified fluid is transferred to a still, and by means of super-heated steam, is distilled at temperatures from 300° to 600° F.; at 450° it yields a paraffine-like solid, called belmontine.

Boussingault gives the following as the result of his analyses of various bitumens:—

	Bitumen of Bechelbronn.	Liquid Bitumen from Hatten, Lower Rhine.	Solid Asphalt, Coxitambo, Peru.	
Carbon, ...	87.0	87.4	87.3	87.4
Hydrogen, . . .	11.1	12.6	9.7	9.7
Oxygen and Azote, .	1.1	0.4	1.7	1.6

The results of the distillation of bitumens will be treated of under the chapter on the products of their distillation.

Naphtha, mineral naphtha, petroleum, rock oil, Seneca oil—under these terms is included a natural product exuding from the strata beneath the soil in many countries; when rectified, it furnishes a transparent volatile liquid, which Dumas considers as a simple compound, while Blanchet and Sell, Pelletier and Walter, and others, look upon it as a compound fluid containing three, if not four, liquids of varying densities. Naphtha contains no oxygen, nor has the liquid any tendency to unite with it under ordinary circumstances.

In Moldavia, Galicia, Lower Austria, in France, England, and other parts of the globe, paraffine substances are occasionally met with, which by chemists and mineralogists are known as ozocerite, earth wax, and fossil paraffine. Hofstädter has pointed out their general affinities; and, according to Hausman, ozocerite has been long used as a candle material as well as paraffine-earth; it is composed of carbon 86, hydrogen 14. *Hatchetine* is isomeric with the foregoing; as also *Middletonite*, examined by Johnston, and found near Leeds, in England.

Peat or turf is the result of the slow decomposition of grass, moss, carices, sphagnum, and other plants which grow in moist situations, where, becoming saturated with water, decomposition goes on slowly and in a different manner from what occurs with vegetable matter exposed to the air. In the case of peat, the bed of vegetable matter does not increase after being first formed, while, in the case of turf, the annual growth of sphagnum, carex, and erica, goes on, and dying down in autumn, adds its vegetable matter to that previously formed, and thus every year a superficial layer of vegetable matter, partly decomposed, is added to the older turf, while the new and annual growth flourishes on the surface.

When freshly cut, peat and turf contain 90 per cent. of water, which by air-drying is often reduced to 60 per cent. The specific gravity varies partly with the amount of water present, but chiefly from the amount of decomposition of the substance; the blacker, denser, and older the peat, the higher is its gravity. Karmarsh found samples of Hanover peat to vary from .113 and .240 in young peat, to .564 and 1.039 in old peat; and of 27 samples of turf examined by Sir R. Kane and Dr. Sullivan, the maximum density was 1.058, and the minimum 0.235, the majority being below .600.

The chemical composition of peat differs considerably

from that of wood: the following samples, from various parts of Europe, show its ultimate composition :—

Locality.	Carbon.	Hydrogen	Oxygen.	Nitrogen.	Analyst.
Vulcaire,	60.40	5.86	33.64		Mulder.
Holland,	59.27	5.41	35.35		"
Philipstown (Ireland),	60.47	6.09	32.54	0.886	Kane and Sullivan.
Tuam "	59.55	5.50	28.41	1.710	Ronalds.

The following table contains the ordinary ultimate composition of the most important varieties of coal and turf: the numbers given are selected from analyses made by Sir Robert Kane :—

	Carbon.	Hydrogen.	Oxygen and Nitrogen.	Ash.	Economic value of 100 parts.
Turf,	58.09	5.93	31.37	4.61	171
Lignite,	71.71	4.85	21.67	1.77	208
Splint Coal,	82.92	6.49	10.86	0.13	262
Cannel Coal,	88.75	5.66	8.04	2.55	260
Cherry Coal,	84.84	5.05	8.43	1.68	259
Coking Coal,	87.95	5.24	5.41	1.40	271
Anthracite,	91.98	3.92	3.16	0.94	278

TABLE,

Presenting at one view the prominent physical and chemical characters of the hydro-carbons derived from woody matters.

Woody Matters.	Hardness.	Specific Gravity.	Fracture.	Lustre.	Color.	Streak.	Texture.	Composition.
Anthracite,	2–2.5	1.8–1.75	Conchoidal.	Vitreous.	Black.	Black.	Brittle.	O
Black Coal,	2–2.5	1.14–1.5	Conchoidal.	Waxy.	Black.	Black.	Brittle, slightly sectile.	$C_{24} H_{12} O$
Brown Coal,	1–2.5	0.5–1.5	Conchoidal.	Waxy.	Brown-black.	Brown.		O
Scheererite,	Soft.	1–1.3	Conchoidal.	Resinous.	Transparent, translucent, white, yellow, green.	Unctuous.		$C^2 H^{11} O_{10}$
Konleinite,		0.83		Pearly, colorless.	White.			$C^2 H$
Fichtelite,					Transparent.			$C^2 H$
Hartite,	1	1.046	Conchoidal.	Fatty, feeble.	White, translucent.	Unctuous.	Not flexible, sectile.	${}^{\{C_2 H_5, \text{ or } C_{40} H_{10}\}}_{\{O H_5\}}$
Ozocerite,	1	0.94–.97	Conchoidal.	Waxy, translucent on edge.	Green, black, yellow, red.	Yellow-white.	Sectile, tough, flexible.	O H
Hatchetine,	1	0.607		Translucent, slightly pearly.	Yellow.			O H
Middletonite,		1.6		Slightly transparent, resinous.	Brown.	Light brown.		
Piauthyrite,		1.115		Opaque.	White.			
*Guaquillite, Irish bog butter,	Soft.	1.099		Resinous.	Bright yellow.	Yellow.	Brittle.	${}^{\{C_{20} H_{17} O\}}_{\{C_{32} H_{22} O_3 + H O\}}$
*Berengillite,			Conchoidal.	Translucent, fatty.	Dark brown.	Yellow-white.	Sectile.	
*Walchowite,	1.5–2.	1.085	Conchoidal.	Resinous.	Yellow, brown.	Yellow.	Sectile.	
*Iolyte,	1–	1.069	Conchoidal.	Very slightly translucent.	Red.	Yellow.		
Piauzite,	1.5	1.008	Imperf. Conchoidal.	Semi-translucent.	Black-brown.	Yellow-brown.		
*Chrismatine,		1.2290			Oil-green, yellow.			
Dopplerite,				Greasy.	Brown-black.	Dull brown.	Elastic on exposure.	$C_8 H_6 O_5$

* Thus marked, are amorphous.

CHAPTER III.

ON THE GENERAL PRINCIPLES INVOLVED IN DESTRUCTIVE DISTILLATION.

BEFORE entering fully upon a description of the various products arising from the distillation of coal at temperatures below that of red heat, it is necessary that something should be known of the changes which occur when animal or vegetable matter is subjected to the action of heat. Experiment has shown that these changes vary in proportion to the range of temperature. In circumstances where the material operated upon is in contact with a plentiful supply of heated air not deprived of its free oxygen by any act of combustion, the whole or much the greater part of the carbon will be burned off as carbonic acid; some carbon may remain behind mixed with the earthy matters of the organic substance forming the ashy coke. The hydrogen in the substance will escape as water at first, and if much be present, as carburetted hydrogen; and with the nitrogen (if the substance contain any) as ammonia, as a carbonate of ammonia. But when the exposure to heat is conducted in close vessels, as in distillation in retorts, another series of changes occurs; at the outset, when the heat is inconsiderable, aqueous vapor, organic acids, ammonia, and some combustible fluids soluble in water, are given off. As the temperature augments, carbonic acid, carbonic oxide, water, and a number of oleaginous substances not soluble in water, are formed. When the temperature rises up to and exceeds a red heat, the products are in great part, or wholly, gaseous.

Destructive distillation may be considered as combustion with a very limited supply of oxygen—merely so much as is contained in the substance itself. The results of the dry distillation of substances vary in so far as they contain or are deficient in nitrogen. Most of the products are common to both conditions, but where nitrogen is an element, there are many substances formed peculiar to it.

In the cases of organic substances not containing nitrogen, as wood, resin, oils and fats, &c., the chief products of distillation are—water, acetic acid, naphtha, or wood spirit, volatile oil, tar, paraffine, creosote, &c.

When the substances contain nitrogen or sulphur, as coal, &c., there are added to the foregoing—ammonia, aniline, leucol, picoline, lutidine, &c., cyanogen and sulpho-cyanogen compounds.

And in all cases, an ashy carbonaceous mass remains in the retort or still, known as *coke*. Such are the changes produced at temperatures below a red heat; at temperatures above this point, a series of gaseous products only are produced: and if there does appear in the ordinary manufacture of gas a large amount of the above *volatile liquids*, it is because they are formed when the retort has from any cause been cooled down below a red heat, when they immediately appear, and are distilled over.

It is the temperature at which coal is carbonized in

close vessels which determines the nature of the products —if it be high, gaseous fluids will be produced ; if it be low, volatile vapors (liquids) alone will form. This is now so well known by manufacturers of gas, that the utmost care is exerted to keep the retorts at a cherry-red heat (1400° F.)

The results of dry distillation are always very complex ; the number of products very great, and difficult of separation.

" The difficulty of tracing the decomposition of organic matters by heat arises from variations in the temperature, and the non-removal of substances already formed, which in turn are themselves decomposed, and the products of two decompositions become mingled together. Thus, under careful management, the distillation of acetic acid gives acetone and carbonic acid ; malic acid gives water, malic acid, and carbonic acid ; but if the temperatures change, another set of decompositions occur, a new set of products are formed, arising from the disruption of the atoms of the first ; thus acetic acid gives marsh gas, and malic acid gives fumaric acid ; hence, if substances be taken, through which, either from their mass or their non-conducting power, the heat cannot be uniformly diffused, a number of different reactions take place in different portions at the same time, according to their respective temperatures ; the bodies generated in the interior are altered as they approach the surface, and hence a very high degree of complexity is given to the ultimate results."—(*Kane.*)

This is exactly the condition in which the distillation of coal is placed ; a mineral having an indifferent conducting power, and in comparatively large masses, exposed to a high temperature, becomes unequally acted on, and frequently, while the interior of the mass is evolving vapors, which subsequently condense into oil, the exterior is giving off gaseous carburets of hydrogen; hence the necessity of having the coal broken into small fragments.

In the production of volatile oils from bituminous substances, the same attention to temperature must be shown; when a dull red heat (about 800°) is obtained, permanent gases begin to form in abundance; hence, on no account should such a temperature be allowed. For all practical purposes, 700° will be found sufficient for the distillation; a temperature of 650° to 700° well kept up, yields the largest results as to the quantity of fluids obtained; and lower temperatures have been found to answer equally well, where super-heated steam thrown upon the materials has been used instead of, or as an adjunct to, external fire.

When substances are composed of only three elements —carbon, hydrogen, and oxygen—or mainly made up of these three, the nature of the change which they will undergo on the application of heat, depends to a great extent upon the ratio which the oxygen and hydrogen bear to each other. When these gases are present in the compound, in the proportion which would constitute water when united, a very low temperature will suffice to draw these elements together to form water, and thus deprive the carbon of its share of oxygen, leaving it as a dry, hard coke. Should the substance, however, be suddenly subjected to a very high temperature, the display of affinities would be somewhat different; for while, at comparatively low temperatures, oxygen prefers to unite with hydrogen rather than with carbon, at higher temperatures, the affinity of oxygen for carbon is more powerful, and thus less water is formed, and more of the oxides of carbon—carbonic oxides or carbonic acid : the hydrogen thus prevented from uniting with the oxygen, seizes an equivalent of carbon, and forms a carbide of hydrogen, which may be either a liquid oil or a permanent gas, according to the temperature at which it is produced.

But if the ratio of hydrogen to oxygen in the material was not in that proportion to form water, but in much larger proportion, then, when heat is applied, the excess of hydrogen seizes on the carbon, and appropriates it, producing the oleaginous or gaseous carbides. Now the proportions of the three elements in which the carbon would be fully saturated with either of the other two, might be expressed thus :—

$$\left.\begin{array}{l}\text{Carbon,} \quad 3 \\ \text{Hydrogen,} \; 3 \\ \text{Oxygen,} \quad 3\end{array}\right\} \text{which by mutual union would form} \left\{\begin{array}{ll}\text{Carbonic acid}= & C\ O_2 \\ \text{Water}= & H\ O \\ \text{Carbide hydrogen}= & C_2\ H_2\end{array}\right.$$

A greater proportion of carbon than that contained in the above relation would not tend to unite with the other elements, but would remain behind as the coke, or carbonaceous residue : when the hydrogen also increases, then carbides of hydrogen, richer in both elements than the formula given above, will be produced, rising to the ratio of $C_{24}\ H_{26}$, and even upwards.

The nature of the products then varies in proportion as oxygen or hydrogen preponderates in a mineral. In the former case, water and carbonic acid will be chiefly formed ; in the latter, hydrogen compounds of carbon will predominate : when these combine at low temperatures, many atoms of each element unite with the other to form definite volatile, oily, and ethereal liquids ; while at high temperatures it is chiefly as gaseous or carburetted hydrogen, it is thrown off; and thus, to manufacturers of mineral oils, the question of suitable temperature for distillation becomes the all-important one.

But no matter how large soever the ratio of hydrogen and oxygen may be in organic substances when distilled in close vessels, the whole of the carbon is never taken up, and hence the residual coke or charcoal is a necessary result of dry distillation.

In this distillation, all of the oxygen compounds pass off first, the richest in oxygen having the start, and followed by others containing less, until the whole oxygen is removed, when the hydrogen compounds then are set free ; thus, in the decomposition of wood in close vessels, the order of production is—water, carbonic acid, acetic acid, carbonic oxide ; oils=compounds of carbon and hydrogen, with little oxygen ; gas=carbon and hydrogen.

In the manufacture of charcoal by the destructive distillation of wood, the object is to get the greatest quantity of carbon left behind ; every circumstance which would tend to remove more carbon than is necessary to unite with the oxygen present, is avoided ; when wood is moist or fresh, when placed in the carbonizing oven or retort, steam is first produced, which, passing over some portion of the wood in strong ignition, tends to be decomposed into oxygen and hydrogen, each of which uniting with some carbon, forms carbide of hydrogen, HC, and carbonic oxide, CO, and thus a less product of charcoal is the result ; dry wood is therefore preferably used. But in the manufacture of oils, the object is not to obtain a large amount of coke, or residual carbon, but the reverse ; hence, every circumstance which would tend to make carbon unite with hydrogen should be adopted. The admission of steam does this in the abstract, but the compounds so produced have not that polyatomic constitution which the mineral oils possess, and hence it cannot be affirmed that the formation of steam in the retort, or its admission during distillation, increases in a *direct manner* the formation of photogenic oils. But of its indirect benefit there can be no doubt, and it, in that case, probably acts partly by keeping the retort at a lower temperature, and partly by assisting to carry off the vapors as they are generated, and by thus relieving the pressure in the retort, tend to hasten the further evolution of the oils.

Organic substances containing much carbon are generally distilled for the following objects :
1. To obtain charcoal.
2. " vinegar.
3. " gas for lighting purposes.
4. " photogenic oils.

And for the economical manufacture of each, ·strict attention is required to the temperature at which the distillation is produced. The 1st and 2d objects are generally effected simultaneously ; but the 3d and 4th cannot be profitably carried out along with the 1st and 2d objects ; nor can Nos. 3 and 4 be carried on together without loss of either class of products—the temperature necessary to produce No. 3 being such as would ultimately decompose any volatile oil produced.

The temperature at which the separation of the constituents of organic substances takes place, involves a large thermometric range, commencing with 300°, and running up to 2732° Fahrenheit.

The tendency of destructive distillation is to produce compounds possessing more simplicity of composition than the original substance, and capable of sustaining the higher temperatures at which they form, unaltered ; so that, under the range of temperature indicated, liquids will be formed when the temperature is least, as at the commencement, and gases when the heat has arisen to the high point set down; and as in the lower ranges where liquids are produced, the effect of augmented heat within this lower range is to lessen the complexity of the compound by dropping or reducing its amount of carbon or of hydrogen, it is at the very lowest temperatures that the liquids containing the highest number of atoms of carbon and hydrogen will be found; and when the temperature arises to that of formation of gas, this gas (a carbide of hydrogen) is produced at the expense of the complex liquids formed at first, which give off some carbide of hydrogen and thus have their proportions simplified. Thus, let $C_{14} H_8$ lose $C_2 H_2$ or one equivalent of olefiant gas, and $C_{12} H_6$ remains; if two proportions of this latter, or $C_{24} H_{12}$, lose six equivalents of carbon by heat, $C_{18} H_{12}$ remains; and if four equivalents of $C_{12} H_6$ or $C_{48} H_{24}$ lose one equivalent of olefiant gas, $C_2 H_2$, one equivalent of marsh gas, $C H$, and 25 equivalents of carbon, there would remain $C_{20} H_{21}$: now the first supposition would show a probable formation of benzole from toluene: the second, the formation of cumene from the doubled atom of benzole: and the third, the formation of paraffin from a quadrupled atom of benzole.

The above, though not perhaps strictly representing the order of decomposition, serves to show the result of augmented temperatures, viz.: the gradual loss of $C H$, and consequently the destruction of the polymeric isomeric hydro-carbons formed at low temperatures; and will, perhaps, also assist in showing, what is desired to be enforced everywhere in this work, that the smallest range of temperature above that necessary to evolve or produce photogenic oils, is sufficient of itself to bring about their destruction.

CHAPTER IV.

ON THE PRODUCTS OBTAINED FROM THE DESTRUCTIVE
DISTILLATION OF COAL.

PEAT, wood, and coal, when subjected to distillation at a red heat, or any temperature sufficiently powerful to destroy the existing condition of the arrangement of their atoms, afford three distinct classes of products, tar, watery fluid, and gas. The proportion of these to each other, and the exact nature of the several products, depends upon the nature of the crude material, and the conditions under which it is distilled. If the decomposition be effected with great rapidity, that is, at a very high red heat, the products will be mostly gaseous—permanently elastic compounds; and the proportion of tar will correspondingly diminish.

The quantity of tar depends upon the two conditions stated, and the proportion of photogenic oils derivable therefrom is dependent, 1st, on the constitution of the crude tar, and 2d, on the temperature at which the second distillation is performed.

When coal is distilled in close vessels, as in the manufacture of gas, heavy volatile vapors are carried over by the heated gas, and deposited in the hydraulic main in the form of tar.

Tarry matters commence to be generated when the temperature rises to 300°, between which and 900° tar is formed most abundantly; when the retort or still exceeds that temperature, tar is formed more sparingly, and when formed, it is in part decomposed within the generating vessel.

Tar is a brownish-black viscous liquid, thickening by exposure to the air, having a peculiar persistent empyreumatic odor. The specific gravity of tar varies from 880° to 975°. That furnished by coal is always the most dense, while turf, schist or slate, and lignite, furnish the lighter tars. That yielded by bituminous schists has the least specific gravity.

Tar almost always has an alkaline reaction; seldom neutral or acid: it solidifies from the presence of paraffin, and absorbs oxygen from the air, the color becoming dark brown (the original color being coffee-brown), and occasionally blackish.

The distillation of tar from coal was first effected at different periods, both in England and other European countries, without any profitable return; in the year 1781, the Earl of Dundonald invented a mode of distilling coal for that purpose, and at the same time to form coke.

The amount of tar derived from close distillation of coal varies, as stated above, with the exhibition of the heat. When the process is conducted slowly, and below 700° F., a large yield is obtained, varying from 16 to 60 gallons and upward, per ton.

The amount has been lately shown to be so dependent on the temperature, that there is little doubt when the latter shall be exhibited judiciously, that 100 gallons per ton will not be considered an unusual amount: one-half that amount is at present looked upon as a large yield.

Peckston, in his history of Gas Lighting, published in 1823, describes the results of the treatment of coal tar thus :—

"When coal tar is distilled in close vessels, it yields an essential oil known by the name of oil of tar; this process requires to be carried on with a very moderate heat." After describing the phenomena occurring in the distillation, he says the oil has the quality of inferior oil of turpentine, and might be used for varnishes; alludes to the residual pitch as resembling asphaltum, and notices the obtaining a lighter fluid (naphtha) by redistilling the oil of tar: from the results of experiments on 50 tons of tar he estimates the product of 1 gallon=$9\frac{1}{4}$ lbs. avoirdupois, as

6.84 lbs. Pitch,
1.26 quarts Oil Tar,
.46 pints Spirits of Tar.

This rate of production may be contrasted with the following results given by Muspratt as the average yielded by the present improved manufacture :—1 ton of coal yields 15 gallons of tar, and two barrels of tar of $4\frac{1}{2}$ each, or 9 cwt., lose by distillation $\frac{1}{4}$th, which is composed of $1\frac{3}{4}$ cwt. and 15 lbs. of essential oil, and 1 quarter and 13 lbs. of water; $6\frac{3}{4}$ cwt. of fatty pitch remains in the retort.

But little was done in the further determining either the constitution of coal tar or its commercial value, until the researches of Laurent and Reichenbach led Mansfield, Selligue, and others, to turn their attention to utilize the

4

liquids derivable from it, for, in fact, the distillation of coal has been considered, until very lately, only in its relation to the production of gas for illuminating purposes, in the manufacture of which, tar was always a certain, and frequently a large product. The composition of this tar has been examined by Runge,[1] Reichenbach,[2] Laurent,[3] Hoffman,[4] Mansfield,[5] Anderson,[6] and Williams.[7]

In examining the results arrived at by these different experimenters, a difference in the substances obtained is found, from which it would appear, as might be expected, that tars formed at different temperatures contain different hydro-carbons.

In the details of the manufacture, more minute results of the amount of production of tar will be given. Where the temperature is above a low red heat, the tar diminishes, and the quantity of tar produced by distillation of coal for gas has been variously estimated, by many superintendents of gas works, at 12½ gallons (282 cubic inches) per ton ; by Peckston, 1½ cwt. per ton ; by Lloyd, 2 cwt. per ton.

From the experiments of Barlow and Wright, the following amounts of tar are produced from the following varieties of coal, per ton :—

	Lbs. weight of Tar.
Pelton Main,	102
New Castle Cannel,	98
Wigan,	248
Youghgelly Cannel,	225

[1] Annal de Poggend., XXXI., 65, 512, and XXXII., 308, 323.
[2] Ibid, XXXI., 497.   [3] Annal de Chim., III., 195.
[4] Ann. der Chem. u. Phar., XLVIII., 1.
[5] Ibid, LXIX., 163, and Quarterly Jour. Chemical Soc., 7.
[6] Philos. Mag., XXXIII., 174.
[7] Chem. Gazette, 1855, p. 401.

TAR FROM CANNEL COAL. 51

	Lbs. weight of Tar.
Boghead,	738
Lismahago,	598
Ramsay Cannel,	295
Derbyshire deep Main,	219
Wemyss,	210

While the foregoing serve as points of comparison by which the bituminous character of the various coals may be estimated, it forms no reliable datum as to the absolute amount of tar which may be extracted from those coals. As this was a bye-product of gas making, in which much of the tar is lost, the figures here are very much below the true product. Allowing 8¼ lbs. as the weight of the gallon of tar, the Boghead coal produced 86 gallons, which is certainly no correct estimate of its real yield under lower temperatures.

The Breckenridge Cannel coal is perhaps the most highly bituminous coal known. The quantity of tarry oil which it yields is 32 lbs. to every 100 lbs. of coal, or nearly ⅓d, or above 820 gallons the long ton: this statement from Silliman's Journal (Vol. XXV, p. 285), appears an over-statement. The Haddock's Cannel coal (Owsley Co.,) yields from 55 to 60 gallons of crude oil to the ton, and from 27 to 30 of purified oil.

The yield of Kentucky coal is given by Dr. Peters, in the 2d report on the Geological Survey of Kentucky, the oil being that produced from 1000 grains of coal:—

	Crude Oil.	Ammoniacal Water.	Coke.	Gas, cubic inches.
Breckenridge Cannel,	318.20	52.10	455.	445
Haddocks Cannel,	248.50	54.50	589.	870
Union Co. mine, bottom part,	148.	88.	750.	465
Mulford's, 5 foot, main coal,	186.50	64.75	684.	567
Muddy River coal,	102.10	119.80	659.50	870
Ice House coal,	108.	73.	714.	465
Youghiogheny coal,	136.	52.	710.	545

Wagenmann examined turf, brown coal, and bituminous slate, to determine the yield of tar. The two samples of turf were from Newmark : coal *A* and *B* from the Mark ; *C* from Prussian Saxony, and the slate from the Rhine country. He obtained the following results :—

1. Firm, dark brown Turf, yielding at 110°, 33.58 per cent. of water, and 6.76 per cent. of ash.—

100 *parts of Turf gave*		100 *parts of this Tar yielded*	
Coke,	27.70	Photogen,	8.90
Ammoniacal liquor,	50.01	Solar oil,	22.56
Tar,	4.89	Solidified paraffin mass,	39.73
Gas and steam,	17.40	Carbonaceous residue,	22.60
		Loss,	6.21

2. Brown Turf, with a fibrous mossy structure, yielding 36.23 per cent. moisture, and 5.49 per cent ash.—

100 *parts of Turf gave*		100 *parts of this Tar yielded*	
Coke,	25.77	Photogen,	7.32
Ammoniacal liquor,	58.03	Solar oil,	21.66
Tar,	5.19	Paraffin mass,	46.03
Gas and steam,	11.11	Carbonaceous matters,	12.77
		Loss,	12.22

3. Brown Coal (*A*), dark brown, firm; sp. gr.=1.369, yielding 29.27 per cent. water, and 7.018 per cent. ash.—

100 *parts of Coal gave*		100 *parts of this Tar gave*	
Coke,	37.66	Photogen,	8.05
Ammonia,	36.69	Solar oil,	45.47
Tar,	5.96	Paraffin mass,	28.52
Gas and steam,	19.96	Charcoal,	13.09
		Loss,	4.87

4. Brown coal (*B*), brown color when dried, breaking readily, with ligneous fibres intermixed, and here and there crystals of sulphate of iron scattered throughout; sp. gr.=1.252, yielding 39.58 per cent. water, and 3.43 per cent. of ash.—

DISTILLATION OF COAL.

100 parts of Coal gave		100 parts Tar yield	
Coke,	30.43	Photogen,	9.10
Ammoniacal solution,	48.41	Solar oil,	38.93
Tar,	4.02	Paraffin mass,	39.43
Gas and steam,	17.17	Carbon,	9.30
		Loss,	3.24

5. Brown coal (*C*), moist, dark brown, masses the size of a large bean; sp. gr.=1.209, yielding 45.258 per cent. water, and 9.83 per cent. of ash.—

100 parts of Coal gave		100 parts of Tar yield	
Coke,	27.36	Photogen,	8.51
Tar,	9.51	Solar oil,	41.48
Water,	49.85	Paraffin mass,	
Sal ammoniac,	0.20	(14 per cent. paraffin)	41.10
Pyrogenic oil,	0.04	Carbon,	5.55
Gas and steam,	13.04	Loss,	3.36

6. Paper coal; sp. gr.=1.264, containing 19.9 per cent. moisture, and 23.52 per cent. ash.—

100 parts of this Slate gave		100 parts of Tar yield	
Residues, with some Carbon=⅓ per cent,	35.69	Photogen,	32.50
		Solar oil,	6.33
Water, some potash, and ammonia,	32.09	Paraffin mass,	51.25
		Charcoal,	8.92
Tar,	25.11	Loss,	1.00
Gas and steam,	7.11		

100 parts of the substances examined yield, therefore—

			Light Oil. Photogen.	Heavy Oil. Solar Oil.	Crude Paraffin.
1.	Turf,		0.435	1.104	1.943
2.	Turf,		0.380	1.124	1.889
3.		*A*	0.480	2.710	1.700
4.	Brown coal,	*B*	0.366	1.565	1.585
5.		*C*	0.810	3.940	3.910
6.	Paper coal,		8.160	1.590	12.870

Paper coal is consequently the most profitable material

for the production of the light oil : yet the use of brown coal, and even turf, is profitable, where the residual coke is a desirable substance to obtain.

Schroeder made an analysis of the bituminous slate from Bruchsal, which yielded in 100 parts—2.5 to 3. per cent. water, 4. to 6. tar, and 100 to 150 c. feet of gas ; from 100 parts tar, 62 parts of liquid volatile oil was distilled, whose boiling point usually ranged between 100° and 350°.

Engelbach, an assistant in the Giessen laboratory, examined the bituminous slate near Bielefeld, which gave 71.20 per cent. of ash. 100 parts yielded—

> 78 parts fixed residues, with charcoal.
> 14 parts watery liquid.
> 1.47 light oil, of sp. gr. 879.
> 1.03 heavy oil, of sp. gr. 955.
> 0.37 butyric fat.
> 0.87 asphaltic fat.

Fresenius made an examination of the brown coal of the Westerwald : as regards the products obtained by dry distillation, 100 parts gave—

Mine.	Variety.	Coke.	Tarry Liquid.	Tar.	Sp. Gr. of Tar.	Gases.
Oranien,	Small coal,	81.97	44.72	5.37	1.043	17.94
Orainen,	Lump coal,	84.86	40.77	3.19	0.952	21.17
Nassau,	Small coal,	81.23	48.69	8.78	1.064	21.29
Nassau,	Lump coal,	81.22	43.07	2.86	1.041	22.80
Oranien,	Lignite, dried,	84.21	42.83	5.61	1.079	17.35
Nassau,	Lignite, air-dried,	86.42		5.88	1.072	12.60

By operating on the crude tarry matters, the following ratio of products were also obtained by Fresenius :

100 parts of air-dried coal yielded—

PRODUCE IN OILS.

Mine.	Variety.	Crude Tar.	Thin Oil.	Thick Oil.	Asphalt.
Oranien,	Small coal,	5.87	1.64	0.41	0.72
Oranien,	Lump coal,	3.19	0.85	0.60	0.44
Nassau,	Lump coal,	3.78	1.54	0.47	0.02
Nassau,	Small coal,	2.86	1.06	0.26	0.51
Both Mines,	Lignite,	5.88	3.01	1.16	1.16

The purification of the oils was effected by treatment with sulphuric acid and bichromate of potash, followed by potass ley, and then another distillation.

Fresenius estimates the yield from 100 parts of crude oil, to be 70 parts pure oil.

P. Wagenmann has communicated the following table, showing the products of distillation of the various raw materials, which yield photogen and paraffin, when examined with the care which the analytic chemist bestows upon such investigations:—

Name.	Locality.	Tar, per cent.	Specific Gravity.	Crude Essence, from 700 to 850 sp. gr.	Crude Oil, from 850 to 900 sp. gr.	Crude Paraffin.
Trinidad pitch,	Trinidad,	70	.875	40	20	$1\frac{1}{2}$
Boghead coal,	Scotland,	83	.860	12	18	$1\frac{1}{4}$
Torbane mineral,	"	81	.861	11	16	$1\frac{1}{4}$
Dorset shale,	Dorsetshire, England,	9	.910	1	6	$\frac{1}{30}$
Rangoon naphtha,	Burmah,	80	.870	50	20	3
Belmar turf,	Ireland,	3	.920	1	1	$\frac{1}{4}$
Georges bitumen,	Neuwied,	29	.865	$8\frac{1}{4}$	14	$1\frac{3}{4}$
Paper coal, No. 1,	Siebengebirge,	20	.890	6	9	$\frac{1}{4}$
" No. 2,	"	15	.880	5	7	$\frac{1}{4}$
" No. 3,	"	11	.880	3	6	$\frac{1}{2}$
" "	Hesse,	25	.890	6	12	1
" "	Rhenish provinces,	11	.880	3	5	$\frac{1}{2}$
" "	Bonn,	4	.930	$\frac{1}{10}$	3	$\frac{1}{2}$
Brown coal,	Saxony (province),	7	.910	2	8	$\frac{1}{4}$
"	Kingdom Saxony,	10	.920	2	4	$\frac{3}{4}$
"	" "	6	.915	$\frac{1}{2}$	4	$\frac{1}{4}$
"	" "	5	.910	$1\frac{1}{2}$	$3\frac{1}{4}$	$\frac{1}{4}$
"	" "	6	.910	$\frac{3}{4}$	$4\frac{1}{4}$	$\frac{1}{2}$
"	" "	$9\frac{1}{4}$	.920	2	5	1
"	" "	6	.910	1	4	$\frac{1}{2}$
"	" "	4	.910	1	2	$\frac{1}{4}$
"	" "	$9\frac{1}{4}$	.920	2	5	$\frac{3}{4}$
"	Thuringen,	5	.918	$1\frac{1}{2}$	1	$\frac{3}{4}$
"	"	5	.920	$\frac{1}{4}$	$3\frac{1}{4}$	$\frac{1}{4}$
"	Neuwied,	$5\frac{1}{4}$	.920	1	5	$\frac{1}{2}$
"	Bohemia,	11	.860	3	5	$\frac{3}{4}$
"	Westerwald,	$5\frac{1}{4}$	.910	$1\frac{1}{2}$	$1\frac{1}{4}$	
"	"	$8\frac{1}{4}$	.910	1	1	
"	Nassau,	4	.910	2	$1\frac{1}{4}$	
"	"	3	.910	1	1	
"	Frankfort,	9	.890	2	6	

Name.	Locality.	Tar, per cent.	Specific Gravity.	Crude Essence, from 700 to 850 sp. gr.	Crude Oil, from 850 to 900 sp. gr.	Crude Paraffin.
Lignite,	Silesia,	3	.890	$1/_{16}$	2	$1/_4$
Lias slate,	Vindee,	14	.870	5	7	1
"	Westphalia,	5	.920	$1^1/_2$	1	$1/_{20}$
Naphtha clay,	Gallicia,	3	.890	1	$1^1/_4$	$1/_{16}$
Turf,	Newmark,	5	.910	1	8	$1/_2$
"	Hanover,	9	.920	$1^7/_{10}$	5	$1/_2$
Black, or Pit coal,	Steier Mark,	8	.890	1	$5^1/_4$	$1/_4$
"	"	6	.890	$1/_2$	4	$1/_4$
White coal,	Australia,	17	.570	6	8	1

Coal Tar is found to contain 3 classes of substances—acids, alkalies, and neutral substances; of the latter class the tar is mainly composed.* These substances may be thus set forth:—

Acids.	Bases.	Neutrals.
Rosalic,	Ammonia, NH_3	Benzin, $C_{12}H_6$
Brunolic,	Aniline, $C_{12}H_7N_2$	Toluene, $C_{14}H_8$
Carbolic, or creosote, $\}C_{12}H_6O_2$	Picoline, $C_{12}H_7N$	Cumene, $C_{18}H_{12}$
	Quinoline (leucol), $\}C_{18}H_7N$	Naphthaline, $C_{20}H_8$
	Parvoline, $C_{18}H_{13}N$	Paranaphthaline, $C_{30}H_{12}$
	Pyredine, $C_{16}H_5N$	Chrysene, $C_{12}H_6$
	Lutidine, $C_{14}H_9N$	Pyrene, $C_{15}H_6$
	Collidine,	Paraffine, $C_{20}H_{21}$
		Ampeline.

There are, in all probability, many other substances not yet sufficiently isolated to be described, which are isomeric with many of the preceding: this is most probably true of the basic substances. The interesting fact about these substances is, as may be seen by inspection of the table, that they all contain only one equivalent of nitrogen, and that, with one or two exceptions, they rise by regular gradations of two of carbon and two of hydrogen, in progressive series, thus—10+5, 12+7, 14+9, 18+13—and so on; besides which, they are all isomeric, or possess exactly the same composition with another series of bases known as the *Aniline* series to chemists.

* Gerhardt, Chem. Organ, Vol. IV, p. 426.

Ammonia exists only in small amount in the tar proper, but the water distilled over with the tar contains the whole of the ammoniacal salts, which can be profitably extracted : the remaining bases are so small in amount, and their properties so little known, that they are objects of chemical curiosity. In a description of the distillation of coals for practical purposes, the consideration of the bases may be therefore passed over.

Of the acids enumerated, but one is worthy of any notice—carbolic acid ; under the name of creosote, this substance has been long known, and widely used for its many valuable properties.

Among the neutral bodies the coal oils belong ; the photogenic liquids derived from the distillation of coals are all enumerated in the above list, which contain three liquid and six solid substances : excluding the latter, the known photogenic liquids in tar are comparatively few in number ; the amount of oils or liquid substances, compared with the solid matter, is, however, so much greater, that the great bulk of the tar is made up of them : of these, perhaps toluene and cumene are the preponderating ingredients.

Future investigations may show that tar contains among its neutral bodies many other constituents than those enumerated : when homologous bodies co-exist in a compound liquid having their specific gravity and boiling points so close to each other, it is a matter of great difficulty to separate each in that purity in which its properties may be examined, and also difficult to state, with accuracy, all of the substances present.

The tars of bitumens, bituminous schists, and turf, contain many of the substances enumerated here, but they also contain fluid oils and solids, in the class of neutral bodies not found in coal tar : hence, the photogenic oils derived from these sources are not always the same chemical substances with those now under consideration, although their photometric value may be precisely the same, dependent upon the proportion of carbon contained in the oil.

The most natural mode of describing the substances produced in distillation would be to take the products in the order in which they appear in the condenser or receiver, on the gradual augmentation of the heat applied ; this method is accordingly adopted.

One of the first products which comes over, in company with a large amount of water, is a mixture of volatile hydro-carbons, which has received the name of crude naphtha, and when further distilled, is known as rectified coal naphtha ; this is further purified by mixing it with ten per cent. of concentrated sulphuric acid, agitating, and setting aside for some hours to rest : when the mixture is cold, five per cent. of peroxide of manganese is added, and the upper portion submitted to distillation. This mode of purification has been recommended by the late Prof. Gregory, of Edinburgh. The specific gravity of the rectified naphtha is 0.850 : it is used extensively as a solvent of caoutchouc, and other allied gums, and also of resins for the preparation of varnish. By repeated purification and fractional distillation, what is termed benzole or benzine, by Pelouze, and others, is obtained: naphtha being a heterogeneous liquid, made up of several hydro-carbons, of which benzine is the most abundant and important.

The numerous applications of which this liquid is susceptible, renders it one of the most valuable substitutes for alcohol, ether, turpentine, and other fluids in common use, as a menstruum for dissolving gums, resins, and other commercial products. Its property of dissolving fat, renders it useful for cleaning cloth, leather, &c., from spots of grease, wax, tar, or resin, without any resulting injury to the color, or permanent odor to the fabric.

Mr. Grace Calvert has pointed out the application of this property in the manufacture of carpets : it had been necessary to oil " slubbing wool " before being spun, and necessary to remove the oil subsequently, so that the color might be more bright ; but this removal was very difficult, and hence the brilliancy of the colors were injured by the presence of the oil, and the carpet soon became faded : but by the use of benzole this oil can be readily removed, and thus the fabric is capable of receiving a brilliant dye.*

When treated with strong nitric acid, benzine produces " nitro-benzole," a substance which is now much used as a substitute for Oil of Bitter Almonds, in perfumery : it is not acted on by ordinary sulphuric acid, but with the anhydrous acid it forms a conjugated acid.

Benzole boils at 186° ; density of the vapor $= 2.38$. At 32° it crystallizes in a gelatinous mass, which melts at 44.6° ; it is insoluble in water, but very soluble in alcohol and ether. On account of its rapid evaporation, Mansfield applied it for the purpose of impregnating gases by passing them through a layer of it ; or by suspending cloths soaked with it in an atmosphere renewed by a fan or blast. The air, when saturated, burns on account of the quantity of vapor present. The evaporation of the benzole, in this process, produces so much cold as, after a time, to check further evaporation ; and hence, this method of producing gas is beset with practical difficulties not yet fully overcome.

* Trans. London Society of Arts.

The formula representing the composition of benzine is $C_{12}H_6$; the substance yields an analysis, in 100 parts—carbon 86, and hydrogen 14. As it contains no oxygen, and, when pure, does not absorb oxygen from the air, it is used to preserve the oxidizable metals, as potassium, &c., from contact with the atmosphere. It yields, when burned, nothing but carbonic acid and water; when sufficient air is not supplied, carburetted hydrogen is produced, and carbon deposited unconsumed.

The light oils of tar which remain, after rectification, on the surface of the water of the main or condenser, are applied, together with the heavy oils, to the preservation of wood from rotting. The permeation of the pores of the wood is effected by placing the latter in close iron tanks, exhausting the air, and then forcing the oil into the interior of the wood by a pressure of 100 to 150 lbs. to the square inch.

These oils are usually toluene, with some cumene, and form a transparent yellow fluid of .820 specific gravity, having the odor peculiar to such distillates; they often contain a good deal of sulphide of carbon : when not separated, the sulphide produces unpleasant results, when used in rooms, by the formation of sulphuric acid.

The following is a summary of the physical and chemical properties of these liquids :—

Toluene was discovered by Pelletier and Walter among the oily products arising from the treatment of resins. Deville obtained it from resin of Tolu, by distillation. Cahours, in the oily liquid which separates from wood spirit, by adding to it ; and Mansfield found it as one of the residues of distillation of coal tar ; he obtained it by rectifying tar, by fractional distillation, and separating that portion which distils between 212° and 382° ; this liquid is washed with half its weight of strong sulphuric acid, and rectified anew. It is a colorless oil, very fluid, not soluble in water, sparingly soluble in alcohol, and more soluble in ether; its odor is similar to benzine; specific gravity=.870; of vapor=3.260; it boils at 237°, (Gerhardt) at 230°. It dissolves in fuming sulphuric acid, and produces a conjugated acid, the sulpho-toluenic acid. Nitric acid transforms it into an oily fluid, nitro-toluene; chlorine acts rapidly on it, forming various chlorides; by oxidation, it is converted into benzoic acid. The formula representing its composition is $C_{14} H_8$.

Nitro-toluene crystallizes, from its hot alcoholic solution, in broad plates: it dissolves in pyroxylic spirit, sulphide of carbon, and the fat and volatile oils in the same degree as in the spirit of wine; it is very sparingly soluble in cold alcohol.

Cumene (cumol) accompanies the foregoing in the coal tars; and in the oil of wood-spirit it is mixed with benzine, xylene, and cymene, from which it is separated by fractional distillation: it is colorless, lighter than water, of a sweet, agreeable odor, and volatilizes unaltered; its boiling point is 314° 5; insoluble in water, but soluble in alcohol, ether, and essential oils; it forms a conjugated acid with sulphuric acid. Nitric acid, in the cold, does not affect it, but on application of heat, a heavy oil, nitro-cumene, is formed. Its formula is $C_{18} H_{12}$.—(*Gerhardt.*)

Both of these oils are highly fluorescent.

These photogenic oils, when pure, should be colorless, and without smell, or with a faintly aromatic odor. Those which smell of creosote always char the wicks, and proportionally with the amount of the impurity. The charring of the wick is consequently a test of an impure oil, or one which contains carbolic acid, as Vohl has distinctly proved. The article sold under the name of double purified coal oil contains 6 to 7 per cent. of creosote. The oil obtained from paper coal on sale in the German towns, contains 10 to 12 per cent. The method of separating the creosote is described further on.

These two oils are, as has been already stated, the valuable photogenic oils, and form the great bulk of the product. It is not possible to state, *à priori*, how much of each of these are present in any coal oil, as it depends on the temperature at which they are distilled. These oils commence to come over with the last portions of naphtha (benzole), and they continue to be distilled until the temperature approaches 400°. As the boiling point of toluene is 237°, and that of cumene 314°, the first portions of the light oil will be chiefly toluene, and the last portions cumene, and if the distillation be conducted from the outset at a very high temperature, but little toluene may be formed. The lighter the oil, the better is it adapted for burning in lamps ; and hence the tar distilled at temperatures not exceeding 320° contain most toluene, while the cumene preponderates when the temperatures is rapidly driven up to and sustained near 400°. This result of a high temperature should be attended to in the manufacture.

Ampeline is a substance resembling creosote, which Laurent has obtained when the distillation runs between 392° and 536°. The crude oil, washed several times with concentrated oil of vitriol, is then mixed with $\frac{1}{15}$ or $\frac{1}{20}$ of its volume of caustic potassa in solution ; allowed to rest for 24 hours, the liquid separates into two layers, of which the lower watery solution is the most abundant : this is drawn off, and agitated with sulphuric acid, which separates an oily liquid lighter than the fluid: this oil is drawn off with a pipette, and treated with water, in which it dissolves, and separates thus any adhering oil; this remaining fluid is ampeline; almost pure, it resembles a fluid fat oil, dissolves in alcohol, and in all proportions in ether; does not solidify at 35° below zero.

The oils which distil over between 340° and 400°, or even 440°, contain creosote. This substance, first described by Reichenbach, is not now generally admitted among the list of true compounds, but, if not identical with carbolic acid, at least contains a large percentage of that substance united with it; it has a specific gravity of 1.037, and boils at 397° 4; on exposure to cold, it does not crystallize; this last property, and its boiling point, are the only differences which exist between it and carbolic acid, and as the other properties and uses of both are alike, the one description will suffice.

They are obtained from the oils by treating the latter with potash, agitating, and distilling the mass; by repeated rectifications with solid potash, the pure liquid is obtained; the potash liquor is treated with an acid, when the impure carbolic acid separates.

It is an oily liquid, highly refractive, fluorescent. Carbolic acid crystallizes by evaporation from its ethereal solution in small prisms, which occasionally melt into liquid at temperatures below which the crystals formed. The crystals melt at 94°. The specific gravity of pure carbolic acid is 1062 to 1065. It is powerfully antiseptic and poisonous, and coagulates albumen; its preserving property is not due to the latter quality; it unites with bases, and forms salts. Sulphuric acid forms a coupled acid with it; nitric acid, chlorine and bromine, form acids with it by substitution.

The liquid with these properties is obtained from coal tar, and, therefore, almost all the substance now found in cumene under the name of creosote, is, in reality, carbolic acid. Wood-tar furnishes the variety of this acid known as creosote.

The composition of carbolic acid is expressed by the formula $C_{12} H_6 O_2$, and may be supposed to be formed from benzole $C_{12} H_6$, by the addition of 2 equivalents of oxygen.

Carbolic Acid, or Creosote, possesses extraordinary antiseptic properties, presenting, to a great extent, the putrefaction of animal substances. Mr. Calvert has used it as a preservative of bodies for dissection, and also to preserve skins of animals intended to be stuffed.

It has been much employed to produce carbazotic acid, by digesting it with nitric acid, aided by heat—a valuable dye-stuff, which gives magnificent straw-colored yellows on silk and woollen fabrics: the acid is easily made pure, and at a moderate cost, and greens as well as yellows are produced, which do not fade. Mr. Calvert has introduced this acid into use. Mr. Bell, of Manchester, surgeon, has used carbazotic acid medicinally as a febrifuge, Mr. Calvert having called his attention to its intense bitter taste, and, in the hands of the former, it has proved a valuable remedy for intermittent fever. Mr. Calvert has also applied it as an agent for preserving tanning matters from undergoing any decomposition by exposure to air, the effect of which is to convert the tannin present into sugar and gallic acid, which results in the destruction of the value of the tanning material, since gallic acid has no tanning properties, and tends even to remove the mordants from the fabric. By adding a small quantity of carbolic acid to the extracts of tanning matter, they may be kept and employed by the dyer as a substitute for the crude tanning material.

That portion of the fluid distilled over at temperatures exceeding 400°, contains but little toluene, and is chiefly cumene. It also contains many of the bases enumerated, some carbolic acid, and a large quantity of paraffine; or if the tar had been made at high heats, naphthaline; to these may be added chrysene and pyrene.

The heavy oil contains a singular organic product, first discovered by Fritsche and Runge, and called by them, "Kyanol," or "Aniline," which possesses the property of giving with bleaching powder, nitric acid, and other re-agents, a magnificent blue color.

Aniline is a colorless fluid, strongly refractive, with a penetrating odor; specific gravity=1.020, and a boiling point of 182°; it dissolves in cold water, alcohol, and ether. Exposed to the air, it absorbs oxygen, becoming yellow and resinous; the blue reaction produced becomes red if acids be added to the solution, and crystals of picrotoxic acid are produced; this reaction distinguishes this base.

The specific gravity and chemical constitution of the light and heavy oils, vary in relation to the temperature at which they were distilled; and perhaps no two distillations give exactly the same relative mixture of the various hydro-carbons of which they are composed; for it must be remembered, as already stated, that *Coal Oils*, as they are termed, are not pure chemical substances, but articles of manufacture; each of the commercial oils containing 2 or 3 of the liquid hydro-carbons, holding in solution small quantities of the solid matters, such as paraffine, naphthaline, chrysene, &c.

The term Coup oil has been applied to the oil obtained

by distilling tar at high temperatures, whereby little, if any, paraffine is produced, the naphthaline being then formed ; the distillation being conducted at 700° F., and the condenser having a temperature between 150° and 175°. The distillate is washed with a hot solution of caustic soda, and afterwards with oil of vitriol ; the clear liquid drawn off is again mixed with caustic soda solution of 25° Beaume. The clear oils drawn off are then distilled in a hemispherical cast iron retort, with a condenser heated to 150° and kept thereat ; distillation goes on until 450° is attained, when a fresh receiver is affixed, and the temperature pushed to 700°. This last oil is washed with soda and acid as before, and again distilled in an iron retort, with 12 lbs. hydrate potass, or soda, mixed with 1 gallon of water for every 100 gallons of oil. The oil which condenses at 450° F. is collected until 650° F. is raised, when the operation is stopped. This oil is Coup oil.

The first oil obtained is what is usually known as dead oil, which contains naphtha, naphthaline, and cymene. Coup oil is not produced by the direct distillation of coal at low temperatures, but always from the secondary distillation of tar at high temperatures, or under conditions that naphthaline may be formed in abundance ; its presence in coup oil prevents the latter from being burned in lamps as paraffine oil is, as the quantity of smoke produced is very great. Coup oil is occasionally formed in the tar of gas works, where the temperature exhibited has been high. Mr. Ross, of England, obtained a patent for making this Coup oil, in May, 1853.

On account of the great variety of constitution in the liquids distilled from coal, it will be unnecessary here to specify their distinct physical properties, as these will be alluded to in describing their commercial manufacture : a slight notice of the characters of the solid neutral compounds, when obtained in a pure and isolated form, will suffice to complete this account of the products of the distillation of coal.

PARAFFINE is always produced by the distillation of organic substances at temperatures below a red heat ; bituminous substances yield the largest amount of paraffine ; but it may be readily obtained by distilling wax with lime. The oil which comes over, solidifies, and the paraffine may be obtained by pressure between folds of bibulous paper. In the distillation of coals, it occurs as one of the last products, concentrating itself in the last portions of the heavy oils, which sometimes become so thick as to solidify below 80°. This constitutes what is commonly called " paraphinized oil," in the language of patent processes.

The paraffine is separated from the oil by cold, and by a centrifugal apparatus, then melted and run into tin moulds, and afterwards subjected to cold pressure first, and finally pressed when warm, and treated with 50 per cent. of oil of vitriol, which destroys the coloring matter, and lastly with a potash lye ; it is then again melted, and run into moulds.

It has great stability—sulphuric acid, chlorine, and nitric acid, below 212°, exert no action upon it. Its property of not being acted on by acids or alkalies, renders it suitable for stoppers for vessels holding such liquids ; also for moulds for galvanoplastic purposes, where the metal is not intended to cover, as a substitute for fat now used.

Paraffine melts at 116° (Regnault), 111° (Kane), and by several experiments made with care at 108°. It boils at 700°, and then begins to undergo decomposition; it dissolves sparingly in alcohol (4 per cent.), but is very soluble in camphene, and in ether, and may be purified by treatment with these last two liquids. It burns in the air with a clear white flame, but requires a draught or large supply of air to prevent sooting; as a candle material, it requires a glass shade to produce complete combustion. It is a ready solvent of some resins, gutta percha, and caoutchouc, with which it unites in all proportions, and destroys its elastic property. As it contains no oxygen, it might be used for the same uses as benzule for preventing oxidizable metals from contact with the air. From not uniting with acids and alkalies it received its name (from *parum affinis*), and this property has been applied to make paraffine paper, for holding caustic alkaline samples. It might also form a tubing substance to transmit caustic gases or vapors. It is too costly, as yet, to supersede white wax, in the manufacture of candles.

Its formula is $C_{20} H_{21}$ in most examinations, but Dr. Anderson states that the composition and properties of paraffine vary with the source from which it is derived, and so of its melting point also.

Filipuzzi examined a sample of paraffine made by Young, in Glasgow, from bituminous slate, which was white, crystalline, without odor or taste, having a specific gravity of .861, at 590° F., and a melting point of 110° at 131° F.; it partially dissolves in alcohol, and separated by cooling. The mass, when separated from the alcohol, and placed under the microscope, showed three different forms, needle crystals, angular grains, and glistening mother-of-pearl scales; by further treatment, he was enabled to separate nine distinct portions, each of which had a different melting point:

Variety—								
1	2	3	4	5	6	7	8	9
Temperature—								
113°	118°	120°	121°	123° 5'	133° 5'	136°	137°	139°

The ultimate analysis of these bodies showed that they were isomeric or polymeric hydro-carbons, viz. :

Melting point—	118°	121°	135° 5'	137°	139°
Constitution— C	85.47	85.93	85.72	85.77	85.69
H	14.29	14.28	14.81	14.21	14.29

By distillation, these yielded a thin, fatty acid, which, treated with potash, sulphuric acid and alcohol, yielded butyric ether. From the experiments made by him, Filipuzzi thinks that paraffine is a derivative of fatty bodies, and is formed from them by some process of reduction.

Dr. Anderson, of Glasgow, who has examined paraffine, states that the products of its distillation are hydrocarbons, radicals of alcohol, density, .750, and boiling at 143° C. Bolley has found that most of the commercial paraffine contains stearic acid : also, that when paraffine is melted it is then readily acted on by chlorine, giving off bubbles of hydrochloric acid gas, and retaining some acid tenaciously. In the compound thus formed some of the hydrogen is replaced by chlorine. It is tolerably soluble in benzine, and the solution may be readily spread upon paper, wood, &c. He suggests the name of chloroffine for this substance.

The lowest melting point of paraffine is given by Laurent as 91°.4 ; the highest, that by Bolley, as 149°.9 F.

Dr. A. found that the melting point of paraffine varies according to the source from which it is obtained. That from Boghead coal melting or crystallizing at 114°, while that from Rangoon naphtha melts at 140°, and that of Turf at 116°. That produced from bituminous coal, by Atwood's process, melts at 110° ; and Dr. Anderson thinks the formula C_{20} H_{21} does not represent the composition of these various paraffines; that the formula C_{20} $H_{20}+H_2$, or more exactly, C_{40} H_{42}; perhaps C_{42} H_{44} and C_{44} H_{46} might embrace some of the varieties.

Naphthalin is a colorless, inflammable solid, crystallizing in plates; it comes over in the receiver mixed with leucol, pyrrhol, kyanol, carbolic, rosalic, and brunolic acids, these forming the oily liquid separated by distillation with water from the pitchy residuum of coal tar. The formula is C_{20} H_8, being the solid which contains the highest quantity of carbon; insoluble in cold, and slightly soluble in boiling water: specific gravity=1.048; of vapor=4.528; it melts at 175°, and boils at 428°, and condenses unaltered in pearly laminæ; it is peculiarly the product of high temperatures, and is yielded by alcohol and organic matters, at a state of high red heat. The crystals of naphthalin may be separated from the impurities by a cold of 14°, and pressure between folds of bibulous paper.—(*Graham.*)

It forms with sulphuric acid, two acids, and with chlorine, yields a series of compounds of great theoretical interest, but of no practical value.

Anthracene is a substance associated with the foregoing in gas tar, and is isomeric with it, the formula being C_{30} H_{12}; it has higher boiling and fusing points, may be distilled unaltered; insoluble in water—copiously in spirit of turpentine.

Para-Naphthalin is polymeric with the foregoing; it melts at 356°, and boils at 392°, subliming in foliated crystals. It is readily acted on by chlorine and nitric acid; its formula is C_{30} H_{12}.

Chrysene and *Pyrene* are two hydro-carbons, first described by Laurent, and are produced in the distillation of resins, as well as in coal. They are among the last products of distillation, when the mass becomes yellowish-red, thick and pasty, clogging the neck of the retort, and containing crystalline plates; on distillation, the pyrene passes over, and the chrysene collects in the neck; they are then easily separable by ether, in which the pyrene dissolves more readily.

To obtain the chrysene, the coloring matter in the neck of the retort is treated with ether, which removes the pyrene, or some oily matters, leaving the chrysene in a pulverulent state: it is crystalline, inodorous, insipid, of a fine yellow color, insoluble in water and alcohol. Ether dissolves it sparingly; spirit of turpentine, boiling, dissolves a greater amount than ether, which is deposited yellow and floreulent on cooling; it melts at 230° to 235° Cent., and on cooling, solidifies into a yellow mass, composed of needle crystals, or thin plates; it distils a little above its boiling point. It is composed of—carbon, 94.7; hydrogen, $5.3=100$, and its formula is $C_{12} H_4$.

Pyrene, $C_{10} H_2$, a white crystalline solid, is associated with the foregoing, than which it is more fusible.

When tar is distilled, a semi-solid mass is left in the still. When the distillate is rectified, a solid pitch or bitumen remains; these are utilized for various purposes.

The carbonaceous mass left at the first distillation, is mixed with the ammoniacal water, and forms a good manure. The tarry residuum of the second distillation is used as asphalt is, for coatings.

There appear to be varieties of coal, which, whether produced by differences in the vegetable species originally composing them, or by different conditions of decomposition, produce different reactions when subjected to dry distillation. It is notorious to practical men, that certain coals yield paraffine at lower temperatures than others, and that some coals produce naphthalin at temperatures which only aid in forming paraffine in the rest.

When the order of decomposition of an organic substance is spoken of, it must be understood only as referring to the exact condition under which it takes place; for under different conditions, a different order of decomposition takes place, and a new set of products are the result: for example, it is usual to speak of coal, that when subjected to a low degree of heat, it decomposes so as to form inelastic or condensible vapors, while, if the heat be augmented, elastic or gaseous products will form; but this is only true of the conditions under which the coal has been treated in that experiment; for, on the large scale, in the operations of nature, we do not find such results to ensue.

The fire-damp which escapes into the galleries of coal mines, leaks out from the fissures and seams in the surrounding coal, and arises from the decomposition of the coal at temperatures but little above that of the atmosphere, but under augmented pressure; the temperature, however, is not that at which volatile liquids or vapors would be produced. The experiments of M. de Marsilly * show that coals heated from 122° F. to 626° F., lost considerable quantities of gas, which began to escape at 212° F., and went on increasing to the limit of temperature attained. The quantity of gas varied from 1 to 2 litres per kilogramme of coal, and toward the close of the operation, from 1 to 2 per cent. of benzine came over. The gas produced was fire-damp, or mono-carburetted hydrogen. This disengagement takes place from all coal

* Comptes Rendus, May 10, 1858.

freshly mined, and is greatest in amount when the coal is finely powdered. It is not obstructed in escape by increased pressure, and after being given off for a time, ceases to be continually produced. The formation and removal of this hydro-carbon gas appears to be the first step in the decomposition of coal, as it appears to go on equally well on atmospheric exposure, or by heating in a retort to 500° F.; it is more rapidly extricated in the latter case, but not more abundant in quantity, absolutely.

The principle which renders coals *fat*, or renders them more coherent, and gives that property to the coke, is perhaps a liquid hydro-carbon very volatile. By exposure for some time to the air, this principle also escapes from coal at common temperatures; or if the coal be submitted to a temperature at 572° F., it also loses this principle, and the coals are no longer *fat;* the coke is powdery and worthless.

That coals may be made to give off elastic gases at low temperatures, is shown by the experiments of Dr. A. A. Hayes,[*] who, by well-contrived operations, prevented the formation of the vapors or liquids which usually are produced; from these the experimenter deduced a theory of the formation of anthracite coal. Whatever force they lend to such a view, the results are interesting, as showing how conditions vary results. In fact, when it is recollected that one of the invariable conditions under which coal is produced, is that of great pressure, it is obvious that the removal of this pressure, as by opening and quarrying a coal seam, and exposing the broken mineral to the air, must be followed by actions of decomposition within the mass, which are furthered and modified but never origi-

[*] Silliman's Amer. Jour. of Science, March, 1859.

nated by the retort, and the furnace of the chemist and manufacturer.

Time plays a less important part than pressure in the production of coal, and therefore less also in its decomposition. M. Barouler[*] planned an apparatus, in which vegetable matters, surrounded by wet clay, and capable of being strongly compressed, could be subjected to a long-sustained temperature ranging from 392° F. to 572° F. The materials were thus placed in conditions similar to that which produces coal, and the apparatus, while partially air and vapor tight, allowed the watery vapor to react on the solid matter under a high pressure.

By placing in the vessel various kinds of wood, Barouler obtained products which, in properties and appearance, resembled ordinary coal, having in places a dull, and in places a brilliant appearance. M. Barouler found these differences to be owing either to the circumstances of the experiment or to the nature of the wood selected for trial, so that in his view, this appeared to explain the formation of striated coals, or those formed of a succession of alternately brilliant and dull coals. He also placed some stems and leaves of plants between the beds of clay, and obtained, at the close of the experiment, only carbonaceous matter, and impressions similar to those found in coal schists.

There is no doubt, however, that the same change which is effected in coals by the dry distillation of the manufacture, occurs also in nature. The occurrence of Ozokerite, Hartite, Middletonite, Fichtelite, and other similar hydro-carbons, show that from coals are formed, by natural processes, bodies isomeric with the paraffine (which itself has many modifications) of the manufacturer.

[*] Comptes Rendus, Feb. 15, 1858.

Many of these substances are found in the seams and fissures of the coal stratum, but the change is still better shown in the liquid bitumens or petroleums, of which that from Burmah or Rangoon naphtha, as it has been termed, is one of the best examples. The raw material is a semi-fluid naphtha, raised from wells sunk close by the river Iriwaddy, in the Burman empire. The geological formations in the neighborhood are sandstone and blue clay. In its raw state, the natives use it as a lamp fuel. The burning fluids which are obtained from it by processes patented by W. de la Rue, are merely separated from the native compound; they are not formed by heat applied, as is the case in the heating of coal, but have been formed by natural processes, and when existing together in variable proportions, constituting the petroleum. When steam at 212° is applied to distil this fluid, several volatile hydro-carbons come over, which require to be separated by subsequent distillations. It is remarkable of these liquids, that though they come over together below 212°, yet, when separated from each other, the boiling points of some of them exceeds 400° F.

It may be remarked, that the presence of hydro-carbon solid resins in organic substances occurs in those which have been subjected to telluric influences for the shortest period, geologically speaking. Fichtelite and Scheererite occur in the latter tertiary, or post pliocene turf of Bavaria. Scheererite, Kenlite, Tekoretin, and Phylloretin, have been found in the tertiary coal of Switzerland;* it is rare to find the congeneric solids in the coal beds of older date, so that we must either suppose that the conditions of decomposition of the older formed coals were not such as could produce those resins—or

* T. E. Clark, Inaug. Diss., Heidelberg, 1857.

what may be more likely, that they were also formed in these, and have been removed by subsequent actions. Their formation in the pine wood of turf-bogs shows that a very low temperature is necessary to produce them, and that moisture and pressure are, perhaps, more actively exciting causes.

CHAPTER V.

ON THE PRODUCTS DERIVED FROM THE DISTILLATION OF SCHISTS AND NATURAL BITUMENS.

WHEN bituminous schists are submitted to destructive distillation, besides the production of naphtha occasionally, and inflammable gas, there is obtained an empyreumatic oil, of a thick consistence. When this tarry oil is submitted to fractional distillation at increasing temperatures, a series of volatile oils are separated, of which the point of ebullition varies between 144° and 540° F.

Laurent gives the composition of those given off at low temperatures, as—

	144°—171°		216°—219°	304°	Average.	
Carbon,	86.0	85.7	86.2	85.60	85.7	$C_2 H_2$
Hydrogen,	14.8	14.1	13.6	14.50	14.3	

Gerhardt remarks, that these oils approach in constitution to tri-carburetted hydrogen.

The oil distilled between 144° and 156°, when rectified with sulphuric acid and caustic potash, is colorless, and has a density of .714, and resembles naphtha in composition and properties, by exposure to sunlight, and to chlorine vapors it forms hydrochloric acid, and thickens.

The oils which are distilled between 360° and 536° F., furnish, by treatment with sulphuric acid and caustic potass, a light, yellow-brown, fatty oil, called *Ampeline*. Soluble in alcohol and ether, and in all proportions with water, it resists congelation 30° below zero. Nitric acid ultimately converts it into oxalic acid.—(*Gerhardt*.)

M. St. Evre, by redistilling the commercial oil distilled from schists, by the fractional method, and purifying them by repeated distillations over potassa and anhydrous phosphoric acid, obtained the following hydrocarbons:—

$C_{26}H_{34}$ boiling between 520° and 536°
$C_{25}H_{26}$ " " 485° and 500°
$C_{24}H_{26}$ " " 414° and 428°
$C_{18}H_{18}$ " " 268° and 275°

The calcareous schists so abundantly distributed over many parts of Europe, are well characterized by the diffusion of bitumen through the mass of the limestone rock.

At Igernay, near Autun; at Gemenval, in Alsace; at Menat, in Auvergne; and in England, in Derbyshire, beds of some extent and thickness have been met with. They have not, however, until lately, been utilized, excepting the schists of Menat, which have been burned to convert into charcoal for decolorizing and disinfecting purposes. The crude distillation of these latter schists furnished—

Oil,	20	
Combustible gases,	{ 14 / 39 }	53 per cent. of combustible matters.
Charcoal and ash,	19	
Water,	8	
	100	

The oil is brown, very fluid, and of a disagreeable odor: in a lamp with a circular wick it burns well, and without smoke, when the diameter and height of the chimney is greater than usual; the flame is brilliant white.

On distilling this oil, and changing the receiver when ⅔ have gone over, an oil comes over, having little color, and depositing crystalline plates, whitish and glittering, when cooled to 32°, or 22°. To separate the crystals, the liquid must be cooled down to 10° C., the whole thrown on a fine linen rag, and subsequent pressure of the crystalline mass between folds of filtering paper. The crystals are further purified by boiling alcohol, whence they are precipitated as it cools. When pure, it fuses at 33° C., very soluble in ether, inattackable by nitric acid, hydrochloric and sulphuric acids, or by chlorine and potassa. Its composition is expressed by the following percentage:

Carbon   = 85.964
Hydrogen = 14.036

—it is therefore paraffine.

The bitumen of Seysell, which is a calcareous rock, yielded, on distillation, according to Dumas:—

Volatile oil, ... 8.6
Charcoal, ... 2.0
Quartz sand, ... 69.0
Calcareous grains, . . 20.4
                        ―――
                        100.0

The bitumen of Bechelbronn is viscid, of a deep brown color, and is used as a lubricating or greasing oil.

The bitumen of Monastier (Haut Loire), does not soften by boiling water, and burns without softening or agglutinating; on distillation, it affords:—

Volatile oil,	7.00
Charcoal,	3.50
Water,	4.30
Gas and vapors,	4.00
Quartz and Mica,	60.00
Ferruginous clay,	21.00
	100.

The bituminous schists do not differ in the products of distillation from pure bitumens: the ashy coke, left as residue, is always more abundant and earthy than in natural bitumens. They have been a long time employed in France, to produce charcoal for decolorizing purposes, due to the fine condition in which the charcoal is left after ignition. The schists of Menat have been long applied to this purpose by M. Bergenhioux. M. Selligue first introduced the manufacture of volatile oil from bituminous schists into France; and operated also with the splint coal of Autun. He obtained these products from the schists—

1. Light or ethereal oil.
2. Fixed oil.
3. Paraffinised oil, used for lubricating.
4. True paraffine.
5. Coloring material, and ammonia.
6. A dry residue, which may be used for discoloring syrups, or disinfecting purposes.

The schists of Vouvaut, in the Vendee, afforded, on analysis:—

Ashes,	61.6
Charcoal,	7.7
Matters volatile at a red heat,	3.2
Oil,	14.5
Water,	3.2
Gas,	9.8
	100.0

In the distillation, water is first given off, then oils, almost colorless, and very light at commencement, deepening in color, and becoming heavy toward the close; density of oil, .870: yields paraffine on cooling.

By fractional distillation of this oil, it yields products boiling at different temperatures. Dumas states that a number of indefinite compounds are thus obtained, and in which no one compound appears to exceed much the proportion in which the others exist. The only practical distinction which can be made in these products, is the division of them into two classes, viz.:—

1st. Those boiling between 105° and 140°—volatile oils.

2d. Those whose boiling point exceeds 428°—fixed oils.

In the distillation of the schists, M. Selligue's chief object was to obtain as much fluid oil as possible, which he applied, before 1845, to lighting purposes, as a substitute for the burning fluids then in use, and also as a substitute for oil, in the production of illuminating gas.

M. Selligue conducts the distillation in cylindrical cast iron retorts, placed vertically; each furnace heats six such cylinders, each of which has the capacity of a cubic metre, and is so constructed, that the schists may be introduced by wagons at the upper part of the cylinder, and the residue drawn off by an iron car run under the lower end. The retorts are so arranged as to economize fuel; the products of distillation are removed from the upper end of the retorts, and are condensed in cooled pipes. When the distillation is one fourth over, the combustible gases produced are turned under the fire-grate, and produce an economy of fuel. The gas is considerable, each cylinder producing 7.500 gallons of gas. Each cubic metre of schist weighs from 1,260 to 1,400 lbs., and yields 90 lbs. of bituminous oil.

From 1 ton of schist, Selligue obtained, in his manufactory, the following products :—

1st. 820 lbs. of light oil, specific gravity 0.760 to .810.
2d. 582 lbs. of mineral oil, adapted to lighting purposes.
3d. 318 lbs. of paraffinized oil, having 14 per cent. paraffine.
4th. 400 lbs. of tar, or residual pitch.

The liquid first obtained is generally naphtha, or benzule, but not constantly nor necessarily so. In this the difference in distillation of schists which are bituminous from true coal or coal shale : the latter always yielding benzule by distillation ; the latter, generally.

Naphtha, if present, will come over at temperatures below 212° ; on heating bitumens to this point, rarely any fluid comes over, as it is only a few bitumens which contain it ready formed. On distillation, they yield petroline when the temperature rises to 450° ; and at a temperature of 482°, the whole petroline distils over.

Prof. A. W. Hoffman examined the bituminous shale of Kimmeridge, Dorsetshire, England, to determine the yield of coke and oily matters, with the following result :

Specimen 1 :—

Coke,	71.5	
Oily Matters,	14.6	{ 2.7 light oil (naphtha). 9.5 heavy oil, containing 1.3 per cent. paraffine. 2.4 pitch.
Gas, water, and ammonia,	13.9	
	100.0	

Specimen 2 :—

Coke,	43.00	
Oily Matters,	39.00	{ 2.3 light oil (naphtha). { 37.7 heavy oil, 1.9 per cent. paraffine.
Gas, water, ammonia, &c.,	18.00	
	100.	

H. Vohl examined the posidonian slate of Wurtemberg, in relation to its capability of yielding oils. 3,000 lbs. of this slate gave, on dry distillation :—

		in 100 parts.
Tar,	289.032	9.63
Ammoniacal liquor,	249.948	8.33
Residues,	2090.505	69.68
Gas,	370.515	12.36
	3000.000	100.00

100 parts of tar (sp. gr.=0.975), yield :—

Photogen,	24.180
Lubricating oil,	41.936
Paraffine,	0.124
Carbon residue,	13.689
Creosote,	19.036
Gas, and Loss,	1.035
	100.000

Theod. Engelbach gives the following percentage, results of the distillation of a bituminous sand from Heide, in Holland :—

Carbonaceous residue,	84.5
Distillate (oils),	14.
Gas, and Loss,	1.5
	100.

The bituminous schists of the United States have

not been examined practically with regard to their productiveness in photogenic liquids. The coal schists of the province of New Brunswick have been treated by the process patented by A. Gesner, in 1854, by which, not more than from 40 to 50 gallons of crude oil per ton were obtained. Shortly after the operation was commenced on the large scale, the Albert coal, or bitumen, was substituted, which, being more easily distilled, led to the abandonment of the schists. The large amount of bituminous coal in the United States will for a long time prevent any attempt being made to distil bituminous schists.

CHAPTER VI.

OF THE PRODUCTS OF DISTILLATION OF PEAT AND WOOD.

WHEN peat or turf is distilled, the chief products are —1st. Pyroligneous acid ; 2d. A brown, empyreumatic, crystallizable oil ; and 3d. Ammonia, and carburetted hydrogen gases. These products are all useful in the arts, and are separated in those countries where peat abounds. A manufactory was erected a few years since, at Athy, in Ireland, for the distillation of peat ; it was worked on the plan in Mr. Rees Reece's patent, sealed in England in 1849. The principle of this mode of distillation is, to drive a current of heated air, and products of combustion, from below, upwards, through the materials in the heated furnace. The heat developed by the products of combustion passing upward, carries off the oils generated. The waste, inflammable products are used for fuel.

The average results of the distillation gave :—

Watery matters, 30.614
Tar, 2.392
Gases, 62.392
Ashes, 4.197

In 100 parts.

The watery products and the tar yielding :—

Ammonia,	0.287
Acetic acid,	0.207
Naphtha,	0.140
Volatile and fixed oils,	1.059
Paraffine,	0.125

The furnace in which the distillation is carried on, somewhat resembles the high-blast iron furnaces; condensers, with scrubbers and main, are attached.

At Denis & Hoeschs, at Ludwigshafen, lignite and turf are the crude materials. The bulk of the latter is reduced by pressure, and subjected to distillation, furnishing a product similar to coal-tar. The turf-tar may be used for purposes similar to those in which birch-tree-tar is used. Turf-coke, as made there, forms a good fuel, and the ash serves for manure.

Turf from Hanover, air-dried, gave, in 100 parts :—

Tar,	9.06
Ammoniacal liquor,	40.00
Coke,	35.32
Gas, and Loss,	15.62

100 parts of tar yielded, on average :—

Light oil (photogen),	19.457—sp. gr.=0.830
Heavy oil (lubricating oil),	19.547—sp. gr.=0.870
Asphalt,	17.194
Paraffine,	3.316
Creosote, and Loss,	40.486

Consequently, 100 parts of the air-dried turf yields :—

Light oil,	1.7633
Heavy oil,	1.7715
Asphalt,	1.5582
Paraffine,	0.3005

Coke,	35.3120
Water,	40.0000
Gas,	15.6250
Creosote, and Loss,	3.6695

The turf *Photogen, or light oil,* is a transparent, light colored, thin liquid, with but a faint odor; it is wholly volatile, and does not become brown by exposure to air; specific gravity=0.835. It is a powerful solvent of fat, resin, and caoutchouc, and after evaporation, leaves them behind unaltered; it contains no oxygen, and has the formula of CH. Burned in camphene lamps, it gives no odor, and when pure, does not char the wick, so that the latter does not require trimming oftener than once in three days.

The nitric acid compounds of turf-oil have an odor of Musk, and Bitter Almond oil, and are used in perfumery and cosmetics. Similar compounds are formed with oil from paper coal and lignite, as well as other hydro-carbons; they all probably contain nitro-benzule.

The oil used alone, or mixed with alcohol, forms an excellent liquid for removing stains and grease-spots.

Heavy Oil, or Grease Oil. This oil is of a clear brown, beer color, has an insignificant odor, and is not so fluid as the turf photogen. Although every good coal-oil lamp burns this oil with a dazzling white light, yet the wick must be cleaned after the lamp has burned from 6 to 8 hours. It has a greater photometric value than the photogen, due to the larger amount of carbon which it contains.

The statement that the light mineral oils are the better materials for lamps, or have a higher photogenic value, is not true.

Mixed with suitable materials, it forms a very good

lubricating oil, which neither becomes resinous, nor hardens by the cold of winter, and is employed for greasing the spindles in cotton factories, in lieu of train or rape oil.

Its specific gravity does not exceed .870, and, like photogen, contains no oxygen.

Asphalt. The pitch obtained after the distillation has a very black color, and is used for varnishes to coat iron-work, and as an ingredient for making lamp-black.

The *Paraffine* produced is very pure, and so large in amount, that it exceeds that produced from any coal, and as a paraffine-producing material, turf has no rival.

The *Creosote* contained in the heavy oil is of a dark brown color, and contains 80 to 85 per cent. of pure creosote; the adulterating ingredients are carbolic acid, butyric and propionic acid, and picamar.

When peat is distilled at an incipient red heat, and gradually augmenting the temperature as the operation proceeds, the tar will contain, besides the volatile and fixed oil, a considerable quantity of paraffine; if the heat passes beyond a certain range, the character of the tar will change, and it will afterward yield very little paraffine.

The works established by the Irish Peat Co., in the county of Kildare, before alluded to, are capable of working up 100 tons of peat per diem. Every ton of peat yields 3 lbs. of paraffine, 2 gallons of volatile oil, adapted for burning, and 1 gallon of fixed oil for lubricating purposes. These are all derived from the tar. The quantity of tar produced by careful distillation varies from 5 to 6 gallons per ton, yielding the above products.

One ton of peat yields 65 gallons of the watery liquor, or nearly in the proportion of 30 per cent. A numerous list of substances have been made out as the products; for all practical purposes, ammonia, acetic acid, and pyroxilic spirit, need only be mentioned.

The fluid from 1 ton of peat affords $5\frac{1}{2}$ lbs. of ammonia, producing, when combined with sulphuric acid, 24 lbs. of sulphate of ammonia. The quantity of acetic acid from 1 ton of peat, is 5 lbs. The naphtha mingled with the water amounts to 8 lbs. per ton of peat. The charcoal or coke left in the retort is equal to 25 per cent. of the weight of the peat.

When wood is submitted to distillation in close vessels, the substances which are produced are very numerous, and differ according to the nature of the wood, and the resinous matters which may be formed by the tree, and contained in its substance. The temperature at which it is distilled, also determines the constitution of many of the products ; which, as in the case of the distillation of other organic matters, are solid, liquid, and gaseous. The gaseous products have been already noticed ; the liquid products are in part soluble, and partly insoluble in water ; the latter forms the *tar*. The liquid matters soluble in water are, pyroligneous (acetic) acid, wood spirit (hydrate of methyle), acetone, and creosote. Those not soluble are the hydro-carbons, toluene, xylite, cumene, and some oxygenated oils. The important solid substance is paraffine.

These distillates are always accompanied with colored, pasty substances, which form the chief bulk of the tar, and which yields ammonia during the distillation, and which, at the close of distillation, becomes a resinous-like substance, which combines readily with alkalies.

When wood-tar is redistilled, there passes over with the watery matters at the commencement, a light-yellow oil, which swims on the surface of the water ; and subsequently, there comes over a thick colored oil, which is heavier than water.

Light Oil. This is a complex mixture, which, when rectified, begins to boil at 158°, but which soon rises by degrees to 482°. The density of these different portions varies between .841 and .877.

To this light oil, Reichenbach gave the name of *Eupion*, under the impression that it was a distinct and unique substance; but Voelckel has shown that the mere volatile portions, those rising below 212°, are chiefly acetate of methyle, with acetone, and a little benzine, xylite, and mesite.

Reichenbach states that pure eupion may be obtained by distilling oil of Colza, having a boiling point of 118°, which rises as high as 336°, in some samples of eupion, dependent on the mode of extraction and the temperature. A substance with such qualities must, as Gerhardt asserts, be a complex mixture of liquids; and it may be asserted that cupion, as a distinct chemical compound, does not exist.

The portions distilling over between 212° and 302°, are chiefly oxide of methyle, as well as the isomeric bodies, benzine, toluene, and xylite; with these are mixed some oxygenated hydro-carbons, from which they may be separated, by washing with concentrated sulphuric acid, which breaks up the latter.

The less volatile portions, boiling between 302° and 392°, are composed of a mixture of hydro-carbons (among which is cumene) and oxygenated oils, separable as above described; capnomore is found among the latter oils.

Heavy Oils. This oil, gathered at the 2d period of distillation of wood tar, is a mixture of some of the foregoing substances, and some other oils heavier than water;

these are attacked by alkalies, and dissolve in such solutions; they are, creosote, capnomore, and pyroxanthogene; the latter, by the action of caustic potass, forms pyroxanthin, discovered by Scanlan, and examined by Gregory; it crystallizes in long yellow needles, and converted by sulphuric acid into a deep yellow-red color.

Besides the foregoing, Reichenbach has described the following substances as derived from wood-tar:—

Pittacal is a substance produced by the action of baryta upon the oil of tar; it dissolves in acids, and is precipitated by alkalies; it does not dissolve in water, alcohol, or ether; it combines readily with alumina, and by its means can be readily precipitated upon the tissues as a dye-stuff.

Picamar is an oil of specific gravity 1.10, greasy to the touch, of a feeble odor, and biting and bitter to the taste; it boils about 518°, and combines with alkalies as creosote does, forming, with them, crystalline compounds.

Creosote has already been described under the products of distillation of coal.

The other compounds do not deserve detailed notice in this work.

CHAPTER VII.

METHODS OF APPLYING HEAT.

THERE is, perhaps, no question of so much moment to the manufacturer of photogenic oils, as that which presents itself to him when about to commence the manufacture—What is the best form and arrangement of the retorts or vessels for distilling the coal or bituminous mineral? All other questions are secondary to this. The mode of purification of the oil, nay, even the selection of the variety of coal to be operated upon, are of less importance than the problem how to obtain, from a given weight of bituminous mineral, all of the volatile and heavy oils which it is susceptible of yielding under the most suitable application of heat.

In the infancy of the dry distillation of coal, where the object was not the manufacture of oils, but either that of coke or of gas, it was deemed desirable to apply a strong heat up to redness, and obtain thereby as much tar as possible; from this tar the oils were afterwards separated by fractional distillation; but as it has been already shown that the nature of tar differs not only with the nature of the substance distilled, but also with that of the heat applied during distillation, the question naturally presents itself, what are the requisite characters which a tar should possess, to extract oil therefrom? or, is it either possible or profitable to obtain oils by distillation in the direct way, without devoting attention to the production of tar as an indispensable necessity? It is now well known, that tars formed at a high temperature, such as that used in the manufacture of gas, yield a considerable amount of naphtha, or benzule, and contain, also, much naphthalin—the proportion of naphthalin being in proportion to the augmented temperature, and the suddenness of its application; but neither of these products are desirable in this manufacture. The substances which are evolved after the naphtha has ceased to be found, and before the naphthalin is produced, being those desirable to be obtained, the point to be gained is, the largest production of them, and the consequent diminished production of the others.

As naphthalin is produced at high temperatures, at which paraffine does not form, and as the production of the valuable oils is accompanied with the slow production of the latter solid; and as the production and distillation of paraffine goes on abundantly at a temperature when the lighter oils have ceased, we have, in this statement of the circumstances pointed out, the conditions of temperature which are necessary to be observed in distillation. The temperature must be above that point at which naphtha or benzule is produced; it must be below that at which naphthalin is formed; within this range, paraffine is produced, and the desirable temperature, therefore, is that between the formation of naphtha and the abundant formation of paraffine.

Paraffine is not produced in tars of gas works, where

a high cherry-red heat has been used; it is naphthalin which is the waxy solid formed. At low-red heats, it begins to be produced; while paraffine is evolved at temperatures beginning at 350°, and running up to a very low-red heat not visible in the day-time; this, then, is the range of temperature within which the manufacture must be conducted. Pouillet has given the following table of temperatures, which will serve as an index to the proper understanding of this subject:—

Incipient redness,	525°
Dull redness,	700°
Cherry-red, commencing,	800°
" brighter,	900°
" full,	1000°
Dark yellow-red,	1100°
Bright ignition,	1200°
White heat,	1300°
Strong white heat,	1400°
Dazzling white heat,	1600°

The range desirable for manufacturing gas lies between 800° and 1000°. That for the manufacture of oils, terminates where that of gas manufacture begins; and perhaps the most efficient temperature is that which does not exceed 700°.

To accomplish the forementioned object, various shapes of retorts have been adopted, where it was deemed desirable to vary the form in present use for the manufacture of gas. In the majority of instances, manufacturers satisfied themselves with the cast iron gas retorts, and accompanying pipes and condensers, and, in some cases, did not even lessen the heat to any considerable extent; it was soon found, however, that the gas thus produced, was at the cost of the oils, and that, with such retorts, the fire must be considerably moderated; and hence, that the direct flame should not be allowed to play on the walls of the retort: this was accomplished either by placing a sole or floor, perforated, between the fire and the retort, or by conducting the flame through flues running alongside the sides and top of the retort.

The ⌒ shaped retort, in such cases, was used, on account of the facility of charging and cleansing; and where horizontal retorts are used, the many advantages this possesses over the circular, or even elliptical ones, may still give it the preference.

The manufacturer should bear in mind that the object he has in view is not the rapid carbonization of the mass, but the slow and gradual one; and hence, those forms of vessels which allow only of slow heating of the mass, are those to be preferred.

In many cases, the coal or bitumen contains considerable sulphur, or pyrites, which, in a short time, so corrodes the inside of iron retorts, as to render their renewal a serious item of expenditure.

Mr. Clegg has shown that clay retorts have many advantages over iron ones; in practice, they have, in Scotland, wholly superseded iron retorts: and are worthy of trial in this country, in those localities near coal deposits, where fire-clay is attainable.

Whatever be the material of which the retort is constructed, two conditions are necessary for success: the first is, that the eduction pipe for carrying off the vapors should be attached to the end least heated, and not, as in gas retorts, from the front; the second is, that such pipes should be of sufficient calibre to allow the free discharge of the vapors formed, and thus no great pressure be exerted within the retort, as such prevents the further formation of vapors; in practice, a pipe less than four inches diameter will not deliver vapors readily.

With regard to clay retorts, it may be observed, that retorts made of Stourbridge clay, or common fire-clay, have been much employed in England and France for the manufacture of gas, and are viewed as being more economical than iron; when made of several pieces, they leak at the outset of employment, but after a couple of weeks' work, they become perfectly gas-tight: they are equally well-adapted for the production of oils, subject to certain conditions of manufacture. The deposit of carbon is always greater in earthen gas retorts than in iron ones, as they are more apt to heat up unequally, and burn portions of the coal inside; this should be carefully avoided in the oil distillation, as the loss of oil is thereby very great. As the material of these retorts is a non-conductor of heat, and apt to be warmed irregularly, the cylindrical-shaped retort will be found best adapted for distilling purposes; the average lengths are, 8 feet long, 14 inches diameter, and 4 inches thick; the mouth-pieces may be cast iron, fitted with bolt and flanch, and jointed with fire-clay and iron cement: from three to five such retorts may be placed over one fire, and the heat should be slowly and gradually increased, and after the operation is finished, they should be cooled equally and slowly: if made of several pieces, the joints may be $2\frac{1}{2}$ to 3 feet long. The cement is made of 20 lbs. of gypsum, made into a pulp with water, add 10 lbs. of iron turnings, saturated with a strong solution of sal ammoniac, mixed to a consistence fit for use. When properly made, they are not liable to fracture from weight of charge, and for economy and durability, are preferable to iron, especially when the coal used is sulphury.

Another mode of applying heat, so as to produce slow distillation, is by the use of *Brick Ovens*. These may be made wholly of fire-brick, or with the bottom and side of fire-tiles, and the crown of brick : some river-sand and pipe-clay, added to the fire-clay of the coal-bed, prevents the brick from cracking.

Mr. Clegg gives the dimensions of an oven as 3 feet 2 inches wide, 8 inches to the springing line of the arch, and thence to crown, 6 inches. The usual charge for such a retort or oven is 5 cwt., and the fuel required for manufacturing gas is estimated at 50 per cent. the amount distilled : as much less heat is required for producing oil, 35 per cent. would be the probable calculation.

These brick retorts are not found as economical as iron for the manufacture of gas, on account of the waste of fuel ; but this would not operate as an objection in the oil manufacture ; and in many localities of the Ohio and Illinois, and Missouri coal-field, where fire-clay is abundant, and cast iron expensive, there is no doubt that the brick oven is the appropriate distilling vessel.

Muspratt, in his valuable work on applied chemistry, under the article Fuel, figures several carbonizing ovens, made of fire-brick, resembling muffle furnaces, in which the materials are placed in sheet-iron cases, or trays, and laid on shelves, or other support, so that the heat may play around them. The muffle has its wall, about $4\frac{1}{2}$ inches thick, heated by a fire, the flame from which circumscribes the whole muffle. The products of combustion pass behind the muffles by a special channel, and return to the front by the flue which leads to the chimney ; the trays may be run in on trucks, to facilitate introduction and removal, by a door in the front, lined with fire-brick, and luted with fire-clay.

The use of chimney and blast-furnaces may be considered an improvement upon the last-mentioned method of applying heat.

The process of carbonizing mineral-fuel for the production of charcoal, involves the use of apparatus which are better adapted for obtaining the mineral oils abundantly, than that used in gas works. When the Irish Peat Company first erected works to obtain not only the charcoal, but also the volatile oils and paraffine therefrom, a furnace resembling somewhat a blast-furnace was used ; closed at the top by a valve-lid, and having an eduction pipe at the upper part of the chimney : these were not found economical, and the furnace patented by Mr. P. M. Crane was adopted. The improvement in this furnace consisted in the use of another furnace to that in which the combustion of the peat was effected. In this additional furnace, the peat is consumed by a blast of air in the ordinary way, and the torrified gases are conducted by a flue to the second one, which is likewise filled with peat, and the heat thus communicated chars the material. Both furnaces may be closed down by weighted valve-lids at top, a, so as to prevent any loss of the products of distillation. The furnace A B is lighted, and the blast of air thrown upon the ignited fuel through 3 tuyeres, c c c, at the bottom : the heated gases are forced over into the furnace C D, and, passing up through the peat, chars it. All of the products, as well those from the com-

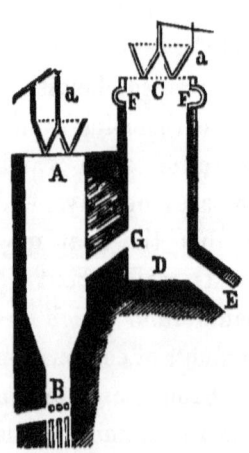

Crane's Apparatus for distilling Coal and Turf, to obtain Oil.

bustion as from the charring furnace, are carried off by the exit pipes F F to the mains and condensers in the usual manner. When the peat is carbonized, the charcoal is drawn out at E into covered tanks or cast-iron boxes.

Next to the form of the retort, the position of it is a matter of importance : as when the retorts are vertically set in the bench, with the eduction pipe at the upper end, as in the apparatus described in the American patents of C. Cherry and R. Schroeder ; these have the advantage of presenting less surface to the fire ; and while the retort is thus kept cooler, it is not so readily corroded, if made of iron ; but, on the other hand, the position of the eduction pipe at the upper part of the apparatus produces much loss by the vapors condensing in the upper part of the retort before reaching this pipe ; these falling back into the retort, strike its hottest portion, and become to a great extent decomposed into gas. The eduction pipes are rarely wide enough ; those adapted for the passage of gas, are much too narrow for oil. Vohl, from his experience at Bonn, states, that where paper coal is distilled, the form of the distilling vessels is essential. Horizontal iron retorts, having wide escape pipes, are preferable to vertical ones, such as are used and recommended in France. The escape pipes should not be above the level of the slate to be distilled, as the oily vapors, having a high specific gravity, rise only a few inches above the surface of the mass, until urged by higher heat, which would decompose the oils.

Delahaye, in France, proposed to use 4 vertical retorts provided with several horizontal pipes from below, upwards. This was a failure. Mehlam, on the Rhine, has a furnace with three retorts, fitted up by Portman & Co. The produce was small, and the consumption of fuel enormous. These retorts are filled from above, and emptied below. The eduction pipes were soon closed by accumulation of dust in them from the cleaning of the retort. The tar obtained in this way was of a dark-brown color, containing 9 to 10 per cent. of coal dust. The yield of oil and paraffine was much less than from horizontal retorts.

Weissman & Co., of Augustine furnace at Bauel, have built a bench of 10 horizontal retorts, two retorts over one fire, and the whole connected by the one main. The retorts are filled in front with air-dried slate, and after the period of distillation, (6·hours,) the residue is removed by iron scrapers. Two retorts were worked at a time, and in one hour's time, so that, in a bench of 10 retorts, in filling the last two were almost exhausted; the watery vapor rising in distillation, carries the oily vapors into the main, and keeps it so warm as to prevent the tar from solidifying. Lime is not necessary for retaining the sulphur.

Flattened ⌒ retorts prove best in operation on paper coal: 8 feet long, 30 inches wide, and 12 to 13 inches high, are the best proportions. The escape pipes should not be too narrow, because each pound of slate gives, along with oil and water, 4 to $4\frac{1}{2}$ cubic feet of gas, at the same time, and this must be allowed to escape rapidly; with these retorts it should be six inches wide. When the pipe issues from the neck of the retort it answers best.

In Herman's furnace, at Gerstingen, in Siegengebirgen, the main is a little sloped, and connected with the escape pipes of the retort by means of coupling-pipes; the produce is received into casks or iron reservoirs; the gas is conducted through serpentine pipes, cooled by water; and it is either used in the furnaces, or conducted into chimnies or blasts for removal.

METALLIC BATHS. 101

The products form 2 strata—the upper, the oil, or tar, and the lower, pyrrhol, with ammoniacal solution, they are separated by means of a bung in the lower part of the vessel.

An ingenious means of distilling at a certain temperature, is that which makes use of the warmth of a mass of melted metal, which always possesses and preserves an uniform temperature.

The process patented by David C. Knab, in England, and dated January 24, 1853, involved the use of baths of fusible metal, which produced the amount of heat required for distillation, at the lowest possible temperatures, by which both the quality of the oil is improved, and the quantity augmented.

In order to obtain the different temperatures necessary, Knab describes several alloys, with fusing points varying from 470° to 797° F., as follows :—

Composition of alloy for Bath.	Melting point.
1. 4 parts Tin, 10 of Lead,	470° F.
2. 4 " 13 "	486°. "
3. 4 " 18 "	505° "
4. 4 " 27 "	525° "
5. 4 " 38 "	542° "
6. 4 " 70 "	558° "
7. Lead alone,	610° "
8. 3 parts Tin, 2 parts Zinc,	612° to 662° "
9. Zinc alone,	680° "
10. Antimony alone,	797° "

These alloys are kept at their melting point during the distillation. The first-mentioned alloy serves for distilling wood, while No. 8 serves for the distillation of turf; asphalte may be distilled by the aid of No. 9 ; while No. 10 is the most appropriate for pit-coal, which, however, yields, with a bath of No. 9, a considerable amount of volatile hydro-carbon.

The apparatus used by Knab consisted of an air-tight metallic box, for holding the alloy, placed above the furnace, through the centre or long axis of the box, ran the retort, which may be of various shapes, according to the material to be distilled; the elliptical retort serves for coals or solid bitumens; the retort is furnished with discharge-pipes for carrying off the volatile products. This apparatus is adapted to the distillation of turf, resins, bones, &c., as well as coal, lignite, and bituminous schists.

Fusible metal has been used as a heating agent, in the direct way, by making the alloy, in a state of fusion, traverse a coil of pipes placed inside of the distilling vessel, using the alloy as steam is commonly used in steam-coils: this is the plan patented by Messrs. Davies, Syers & Humfrey, in England, December, 1855.

In order to obviate the injury to the retorts, arising from the continued application of heat upon the under-surface, movable retorts have been adopted, involving the partial or complete rotation of the retort. One of the earliest forms of movable retorts was that breveted by Gingembre, of France, and described in the Brevets d' Inventions, Vol. IX., p. 235, and pl. 26. In that case, the retort, after having been charged and distilled, was, before the new charge was inserted, turned round one-fourth of a revolution, and thus a new surface of the retort was presented to the fire; when again in operation, another turn was given, and thus, after four distillations, the same surface came again over the flame. By this means, the iron retorts were made to last five or six times as long as they otherwise would.

The advantage of this change of surface, led to a motion of revolution going on during the operation of distillation. The brevet of Beslay & Rouen, described in Brevets d' Inventions, Vol. XLIII., pl. 6, patented this mode of renewing the surface over the fire.

In April 15, 1858, David Alter & S. E. Hill, obtained a patent for an improvement in distillation of oils from coal, &c., which consisted in the use of a revolving cylindrical metallic retort, with the eduction-pipe coinciding with the axle, and connected with the condenser; the motion was communicated by a series of wheels worked by a weight, or other power. The retorts revolve slowly.

T. D. Sargent obtained a patent in June 15, 1858, for a revolving retort, made of clay, and worked somewhat similar to the foregoing; other patents, involving slight novelties in the internal arrangement of the retort and issue-pipes, have been obtained within the last two years.

The use of revolving-retorts, while leading to economy of the iron-retort, is perhaps no economy in the general manufacture; for, accompanying the rotation of the cylinder on its axis, is the constant disturbance of the coal, which, while it presents fresh surfaces of the mineral to the action of heat, also, by attrition, grinds into a powder, a portion of which is always necessarily carried up with the fluids, and gives not only a dark-colored, thick oil, containing solid matter undissolved in it, but the pipes and eduction-tube are very apt to get clogged from the accumulation at the bends of the apparatus. The exit-pipes are rarely wide enough, in this form of retort, to meet or cope with this accident. The oil obtained requires additional purification.

To prevent this fine dust being carried over the end of the eduction pipe in the retort, is, in some apparatus (as in Sargent's), bent upward at an elbow to reach nearly the inner surface of the cylinder; in others, as in that patented by Jas. Gillespie, March, 1859, the mouth of the pipe is hopper-shaped, and is kept in a stationary vertical position inside of the retort, by means of pins, which surround and are inserted into the journal.

In the apparatus patented by J. & W. B. McCue, the retort revolves $\frac{1}{4}$ of its periphery by wheel and crank arrangement, and then returned to its original position, and this motion is repeated several times during distillation; the interior of the retort has ribs running along its inner surface from front to rear, and a few inches apart; by this means, the coal was somewhat kept in place, and the constant agitation modified to a lesser degree.

As all revolving retorts must be built in loosely in the brick casing, they are liable to get out of place, and the mechanism of operation is much more complex; these circumstances are drawbacks to their operation and extension in use, were they not subject to the objections already stated.

There is no evidence yet afforded to show that the use of revolving retorts is accompanied with any solid advantages over stationary retorts.

The coal or bitumen in the retort being very apt to burn, many attempts have been made to prevent such an accident, the most obvious plan being, to keep the coals in motion. The revolving retorts effect this to a degree. Additional apparatus for stirring has been recommended. Solid arms traversing the retort in its long axis, having blades or stirrers attached thereto, have been patented in England, in 1854, by Astley P. Price, who used a vertical retort, with agitating arms revolving in the centre, and allowing the coal to fall down at the sides; the coal is fed in by a hopper, and an Archimedes screw used sometimes instead, placing the retort at variable angles of inclination.

The English patent of F. Archer and W. Papineau, dated December 15, 1854, involved the use of feed-rakes, or agitators, which, on revolution, made a screw-like motion, and thus pushed the materials progressively onward. A horizontal cylindrical retort, fed by a hopper, was the apparatus used in this patent, in combination with the feed-rakes.

In this country, similar agitating knives or stirrers, are described in the apparatus of C. Cherry (1856); John Nicholson (1859); Joseph E. Holmes (1859); and N. B. Hatch (1859). The use of agitators is open to the same objection as that of revolving retorts, viz., raising a quantity of fine dust, which is carried up and clogs the pipes: this acts to a less extent where the retort is vertical, than when it lies horizontal.

The Archimedean screw is an old application in gas retorts, and has been transferred to coal-oil stills. The apparatus patented by Count de Hompesch contained this screw, which, while it kept the coal in agitation, also delivered the coke out of the farther end of the retort; the screw, in this case, fitting the internal bore of the retort accurately. The retort was placed horizontally, with a slight slope.

In the American patent issued to Jas. O'Hara, Feb., 1859, an upright retort is worked with an Archimedean screw of less diameter than the bore, the result of which is, that, while the coal in the centre of the retort is gradually drawn upwards by the screw, the mineral lying between the edge of the screw and the wall of the retort is in a state of continual descent, and thus a constantly progressive motion of the material distilled is being kept up.

In order to relieve the interior of the retorts from the pressure of the vapors as they are generating, various means have been adopted to remove the nascent vapors Aspirators have been used in gas works for a similar object, and their application in this species of distillation has been attended with great benefit.

Aspirators have been used in the processes patented by Bellford, and others, in England, and by L. Atwood, in this country.

A. E. L. Belford, in his English patent (August 22, 1853), describes the vertical tower or furnace, whose height is four times its breadth, and capable of holding three tons; when the charge was delivered in by a manhole above, a fire was lighted below, and the distillation allowed to proceed. The heated products of combustion passing upwards, carried upward the oils formed by the action of the fire upon the stratum of coal immediately above it, which, as the fire progressed, was driven through the exit pipe at the upper part of the chimney, into the condenser, the coal, as it was consumed, gradually dropping downward, and ultimately being removed by rakes through the opening immediately above the fire-bars.

Wm. Brown patented, in England (August 23, 1853), a mode of distilling coals and bitumen in an upright tower, making use of the fire beneath to volatilize the newly-formed products.

It may be observed of the distillation by towers or chimneys, that two principles, distinct from those in common use, and not hitherto described, are involved; the first being the absence of application of external heat to the distilling vessel; and the second, the distillation by means of gaseous or vaporous matter, not in a state of ignition : each of these principles deserve some notice in this place.

It has been already frequently mentioned, that solid substances, exposed, in close vessels, to naked flame, are very liable to become burned in some places, and to become converted, at such points, into gas, while, in other parts of the same vessel, the distillation of condensible vapors is going on slowly ; in iron retorts, this is constantly the case, even when it is coated internally with an enamel or glaze, as in the apparatus patented by Messrs. Evans, in England, Sept. 14, 1854. To avoid this burning, heating the mass from within has been tried, and forms the subject of many patents.

Open steam, or steam of ordinary temperature and pressure, has been used in various ways, either by admitting it by a perforated pipe into the retort, while the outside was heated in the ordinary way ; or by wetting the coal before being placed in the retort ; as the temperature of the retort rises, the water becomes vaporized, and carries the oils over with it. Used in this way, steam cannot be considered properly a heating agent, since its chief action is to absorb heat from the coal, and thus really to keep the temperature of the inside of the retort below what it would otherwise be. But it is otherwise when steam is used either under high pressure or as super-heated steam ; in the latter case, its action is twofold—it raises the temperature of the inside of the retort ; and 2dly, it decomposes the coal into oils. The steam is generally super-heated by being passed through coils of iron pipe traversing the furnace beneath the retort : where external heat is not applied to the retort, a separate fire is required. When the steam is heated to 660°, or thereabouts, distillation of the coal goes on freely, and nc external heat is needed : but in very many patented processes, both internal and external heat are applied contemporaneously ; such is the case in the patent of Wm. Brown, January 13, 1853 ; also of J. F. F. Challeton, October 21, 1853 ; of G. F. Wilson, February 15, 1854 ; and of J. Chisholm, December 19, 1853 ; in all of which, except that of Wilson, super-heated steam was used internally.

Super-heated steam is one of the most powerful agents in the hands of the chemist for producing chemical decomposition ; and the happy applications of Violette and Scharling to various technical purposes, are proofs of its advantageous use in many cases.

In the distillation of coals or other materials to produce hydro-carbon oils, it may be questioned, however, whether its use is advisable ; for, if we consider that the object of destructive distillation in close vessels is to preserve the material from the oxidation of external air, which, at high temperatures, is very powerful, the use of a substance capable of yielding, and which does yield oxygen abundantly at high temperatures, may very properly be considered inappropriate : at high temperatures, carbon has sufficient affinity for oxygen to abstract the latter from hydrogen ; and hence, steam is decomposed by charcoal, and carbonic oxide produced with the formation of hydrogen. This liability of the carbon to be oxidated is to be obviated if possible ; and when heated gases or vapors are to be used, those should be selected which do not yield oxygen : air, whose oxygen has been wholly converted into carbonic acid by complete combustion, is such ; but it is difficult, if not impossible, practically to obtain any supply of it. Nitrogen gas, derived from the deoxidation of air, would be a valuable agent ; perhaps the most efficient would be hydrogen gas, which, by uniting with the free carbon in the retort, might itself increase the production of oils; hydrogen might be obtained by the decomposition of water in redhot vessels in the presence of scraps of iron, &c. It is, however, difficult to apply any gas or vapor with sufficient economy to make it a useful improvement; and where it has been once used, as by Wagenmann, the practice has been given up as being too complex and too little remunerative.

The only exception to this, is, the use of the heated products of combustion, in which the air is, to a large extent, deprived of its free oxygen by the formation of carbonic acid and carbonic oxide; there is present, also, some water in the form of steam, and some sulphuric acid, where the air has previously passed through ignited coal.

The only positively deleterious agent present, is that portion of heated air which has not lost its free oxygen. This proportion is, however, small; and therefore, of all means of heating by gases passed through the material itself, the "heated products of combustion," as these gases are called, is the most economical and desirable.

In the English patent of Wm. Little, dated Feb. 4, 1854, these products are employed as follows: air is either driven through a fire by a blast, or drawn through by an aspirator; after being "burned," as it is termed, it is passed into the distilling vessel at its bottom, which it warms *in transitu;* it then passes upwards through the coal, to be discharged by an eduction pipe or chimney at top, associated with the hydro-carbon vapors which it produced in its passage.

This is an upward distillation, and is liable to the objection already raised against the method of ascensional distillation for obtaining hydro-carbon oils, namely, that the oils, when formed, are readily condensible in the upper part of the still, and tend to drop down into the bottom of the heated retort, and become converted into permanent gases.

There is no question that the adoption of distillation by gaseous heating internally, and the removal of external heat by naked fire, leads to a large increase in the produce of oil—an increase from 33 to 60 per cent. in many cases —that is to say, that coal yielding 33 per cent. of crude oil by naked fire, may be made to yield even 60 per cent. by using burned air alone. When the merits of this process become better known, it will be more universally adopted.

The principle involved in the method of distilling coal, patented by Dr. L. Atwood in October, 1858, is virtually that by which the peat is distilled in the apparatus of Crane, already referred to; but in carrying out the principle, the practice is modified. In both, the distillation is carried on by the heated products of combustion, being passed through the material to be distilled, no external fire being applied; the retort has a tower or chimney-shape ("pipe," technically), into which the coal is delivered from above; a fire being lighted at the top of the chimney, the eduction pipe is placed at the bottom, and is furnished, before it reaches the condenser, with a steam jet-pipe, which, when operated with a steam blast, by exhausting the tower, determines a current downward, carrying the heated products of combustion (carbonic oxide, carbonic acid, some steam, and some undeoxided air) down through the coal, and thus distilling the oil which it carries along with it into the condenser.

The progress of distillation is more constant and uniform in this tower retort than in any other form of apparatus. The tower has many advantages : 1st. Its great capacity, which is now being made capable of holding 25 tons in one of Atwood's pipes ; from one to four tons are the more ordinary quantity distilled in a tower ; the distillation goes on until the oil ceases to come over, and is usually three days in operation : when it terminates, a jet of Croton water from a hose is made to play on the upper end of the tower until the coal is extinguished. Fire-brick and fire-stone are the materials.

The advantage of Atwood's mode of distillation over the other analogous processes, which involve the use of the heated products of combustion as the agent of distillation, consists in conducting the vapors downwards. By this means, the nascent oils do not by condensation fall back or down upon the fire beneath, and by being converted into gas, cause a loss of the distillate: this is what must occur in the method adopted by Du Buisson, which, in almost every other respect, resembles Atwood's.

Dr. Atwood has patented other forms of apparatus for distillation of coal oils, all, however, preserving this feature of downward distillation : thus, in one apparatus, the fire is external to the tower, and communicated with its upper part by a series of flues, through which the heated gases are drawn into the pipe.

From the foregoing details of the various modes of applying heat, it is evident that the improvements made have been in lessening the actual amount of heat applied, in distributing the heat more equally over the whole inside of the vessel. In the gradual desuetude of the practice of external heat to retorts, and in the use of chimney retorts or pipes, and in the use of the heated products of combustion, as the agent for supplying the internal temperature in the pipe. Viewed independent of local causes, which occasionally determine certain methods of manufacture, not in themselves generally desirable, the process, as patented by Atwood, may be looked upon theoretically as *par excellence* the most advantageous method of distilling photogenic oils.

CHAPTER VIII.

COMMERCIAL MANUFACTURE.

IN another portion of this work, the result of the distillation of coals and bituminous substances, with regard to the amount of produce obtained, has been given. Inasmuch as the bituminous differs in no respect intrinsically in whatever country found, the results of the distillation of American coal can in no respect greatly vary from that of European coals : the difference in the results are due to the different modes of working, as it has been shown in the chapter upon the application of heat, that the application and continued exhibition of the appropriate temperature has more effect in producing abundance of oils, than even the different quality of the coal, where the variability of the latter is not extreme.

The following table of results of examinations of Cannel coal from different localities, has been kindly placed at the author's disposal by Prof. Asahel K. Eaton, of New York :—

State.	Locality.	Amount of Crude Oil per Ton.	Ammoniacal liquors, per cent.
Kentucky,	Breckenridge Cannel,	140 Gall.	83
"	" "	100 "	83
Virginia,	Cannelton,	105 "	16
"	near Cannelton,	93 "	20
Ohio,	Cochocton Co.,	87 "	20
"	"	60 "	25
"	Mahoning Co.,		
"	Lower Bed, Canfield,	75 "	83
"	Middle " "	70 "	83
"	Upper " "	60 "	83
"	Mahoning Co., Lima, Upper and Middle Bed,	70 "	20
"	" Lower Bed,	45 "	25
"	" unknown locality,	66 "	83
"	Jefferson Co., near Steubenville,	70 "	80
"	" " "	45 "	16
"	Columbiana Co., E. Palestine,	45 "	25
Pennsylvania,	Beaver Co., Darlington,	55 "	83
"	" near "	40 "	25

The specific gravity of the crude oil, as it runs from the retorts at a temperature of 60° F., was about 21° Baumé.

In contrasting the results which such tables as the foregoing yield, with those obtained by commercial manufacture, allowance must be made, on the one hand, for the delicacy of the manipulation, and on the other, for the non-attendance to the precautions needed to avoid loss in distillation; this loss, or the difference between the results, amounts to nearly 25 per cent.

Professor Eaton believes that the loss in actual working results arises from inattention to the following indispensable conditions:—

1st. That coal should be finely crushed, the finer the better.

2d. The retorts should be worked at a very low red heat—the retort not being visibly red by daylight.

3d. There should be no pressure whatever upon the retorts, but, if possible, a slight exhaust action.

4th. A gas-tight retort. Leakage of the retorts accounts for much of the difference between practical results and laboratory practice, a difference which will cease to be found whenever the above conditions are regarded by the manufacturer.

The produce by the operation of the revolving retort, patented by Alter & Hill, is very great. The retorts are now used 8 feet long, by 6 feet in diameter; the retorts are charged every 2 hours with 16 bushels of the cannel coal of the vicinity,* and the charge is renewed 12 times daily; the yield is, on average, 500 gallons daily of crude oil, which, on purification by simple redistillation in large iron stills, yields about 70 per cent. of commercial coal oil.

The produce of the Lucesco and other works will be given farther on, when treating of the localities where the manufacture is carried on.

According to Dr. Augustein, in 1855 there were three establishments in Germany, one in Hamburg, that of Wiesmann & Co., near Bonn, and that of Denis & Hoech, at Ludwigshafen. The number since then has increased.

The Hamburg establishment uses cannel coal, and treats the distillate with sulphuric acid after it has been distilled several times; it then resembles rock oil or petroleum, having a specific gravity of .785, and having very little of the peculiar odor of the mineral oil, being very free from sulphur, and is very superior to other similar products on that account, thereby allowing its use in the worst ventilated rooms, and has a photometric value, compared with oil, as 4 to 1.

These oils are much used for street illumination in Northern Germany; at the Hanover railroads, in all lamps placed out-doors, for which it is well adapted, as it never freezes during winter.

* Thirty miles above Pittsburg, on the Alleghany river.

The residuum of the second distillation is used in the manufacture of the artificial fuel, known as the Charbon de Paris.

Paraffine is not prepared in Hamburg. The residual paraffinized oil is broken up or subjected to another distillation at high temperatures, in order to obtain light fluids by the decomposition of the paraffine.

At the establishment near Bonn, lignite, found in that vicinity, called leaf or paper coal, is operated on, which is distilled at a low red heat, in iron retorts like those used in gas works; a blackish tar and ammoniacal liquid are the products. The former yields 90 per cent. of oils, 50 per cent. of which are thin enough to burn in lamps; they are purified by treatment with sulphuric acid and alkaline ley.

The process of manufacture of the hydro-carbon oils, is carried on at Bonn by Mr. P. Wagenmann thus:—

The bituminous coal is broken into small lumps the size of a nut, and where the coal contains sulphur, it is sprinkled over with milk of lime; the coal is then placed in a desiccating furnace, 60 metres long, 6 metres wide, and divided into compartments with high walls more than half a metre high, and $1\frac{20}{100}$ metres * high; these form the supports of so many vaults on which the schists are placed to be dried; below these are the waste residues of the distilled materials.

When the coals are dried, they are distilled in retorts resembling those employed in manufacture of gas, except that the exit pipe is at the end opposite to the mouth; two retorts are set over each fire: the retorts are about 7½ feet long, and 22 inches wide, with discharge tubes 12 to 15 centimetres wide. The flame plays only on

* The metre is equal to $39.\frac{37}{100}$ inches.

the bottom of the retorts, and thence passes into the chimney.

Wagenmann prefers a bench of 16 retorts and 8 fires, disposed round a central chimney, so that the flame may circulate from one flue to another, and submit the retorts to an increasing heat: the products of distillation of the 16 retorts are led off by an iron pipe 78 feet long and 23 inches wide, kept constantly cooled by a stream of water on the outside. When the gases are passed through this pipe, they enter into a large iron cylinder filled with coke, which rids them of the last traces of tar they may possess; thence they flow into a chimney 14 feet high, furnished with a draught and regulator.

The liquid products of distillation flow into a grand reservoir, kept constantly at a temperature of 30 C., in which the tar separates from the ammoniacal waters; these waters are mixed with the residue of the large retorts, and furnish an excellent manure.

The tar is drawn up by pumps into the purifying apparatus, where it is mixed with sulphate of iron in the proportion of 1,000 parts of tar with 40 parts of sulphate digested together for three-quarters of an hour at 30° cent.* The purifiers are large cast iron vessels of the capacity of 20 hectolitres,† in which the iron pipes move by mechanical power.

The tar thus freed from sulphuret of ammonium is introduced into distilling vessels capable of holding 350 gallons, and distilled by superheated steam. The products of distillation condense in a leaden coil 32 to 39 feet long, and 7 to 8 cent. wide. The following products of distillation are distilled in this way, i. e., by fractional distillation:—

* Centimetre—$\frac{39}{100}$ of an inch.
† A hectolitre is nearly equal to 26¼ gallons, wine measure.

1st. Volatile liquid, having specific gravity=.700 to .865
2d. Heavy oils, for lubrication,   "   =.865 to .900
3d. Paraffine,   "   =.900 to .930

These three substances are treated each with 4, 6, and 8 per cent. of sulphuric acid, 1½ and 2 per cent. hydrochloric acid, and 1 per cent. acid chromate potass, with which they are agitated for half an hour. Allowed to rest for three hours, they are poured off the dregs, and mixed respectively with 2, 3, and 4 per cent. of a ley of caustic potass (marking 50 Baumé) in iron vessels: finally, each of the products purified are placed in a still, and distilled by superheated steam.

No. 1, mixed with No. 2, so as to obtain the specific gravity of 0.820, produces the *Mineral Oil*, or *Photogen*, which is burned in lamps adapted for that object.

Part of the product distilled from No. 2 having a specific gravity .860 to .700, forms the *Solar Oil*, which may be burned in Argand or Carcel lamps.

The remainder of No. 2, mixed with some of the product of No. 3, furnishes the *Lubricating Oil* for machines.

The rest of No. 3 is introduced into a vat, where the temperature is lowered until it crystallizes: in three or four weeks, the paraffine crystallizes in large tablets, and is separated from adherent oil by centrifugal machines making 2,000 revolutions per minute; the paraffine, melted and rolled into squares, is submitted, while cold, to the hydraulic press, under a pressure of 300,000 lbs.: it is melted again, and treated with 50 per cent. of concentrated sulphuric acid at a temperature of 180° C. At the end of two hours, the paraffine separates from the acid, and is washed with water; it is then run into cakes, and pressed, while hot, between two layers of hair cloth in the hydraulic press, melted anew, and mixed with 5 per cent. of stearine, for some hours, at a temperature of 150° C., in a leaden apparatus, and finally mixed with 1 per cent. of solution of caustic potass, marking 40° Baumé. At the end of two hours, all impurities are precipitated, and the paraffine, limpid as water, is ready to be drawn off.

Wagenmann, having worked one year with the process just described, which was patented by him in 1853, found it very defective in operation ; the stills were unreliable, owing to unequal action of the heat, causing—1stly. Irregular results in distillation; the adjustment of the heat not being manageable, so as to keep down the augmenting temperature of the oil. 2dly. The time occupied in distillation was too long—the oils requiring two rectifications ; one still containing 1,500 quarts requiring 36 hours for distillation, and 12 hours for subsequent cooling and purifying, there could then be only two distillations of 1,500 quarts, in 96 hours. 3dly. The separation of the oils was very incomplete ; if a still, distilling oil of .870 specific gravity, be put out of operation, allowed to cool, and then repeated, it will distil oil not of .870, but of .920 specific gravity : for this reason, the oils so distilled furnish always but little of a light thin fluid, and contain paraffine, which is not desirable. 4thly. The temperature in the still, even when steam was employed, became too high, the loss also being large, as only 91 to 92 per cent. of distilled fluids are obtained from the still. This loss was so great as to have led Wagenmann to adopt a different mode of distillation ;—he was led to think that distillation in vacuo would remove all these defects.

To remedy the injury arising from suddenly cooling or heating iron retorts, he added ends of copper to the iron body of the retort, and used copper riveting. The distillation of tar, after it is separated from sulphide of ammonium, commences above the temperature of high pressure steam, and is best effected by the combustion of the gas, derived from the distillation of the crude tar.

The apparatus consists of two sections of a sphere, with a cylindrical-shaped vessel in the centre, capable of containing 1,500 to 1,800 quarts, with a diameter of 6 feet; the lower hemisphere is surrounded with a jacket perforated with holes, opening into a pipe leading to the flue; the gas burners enter through apertures in the lower part of the jacket, the gas burners consuming 80 cubic feet per hour; a try-cock is attached to the lower part of the vessel; also one for the admission of steam, through a circular coil; on the cylinder is the cock connected with the supply-box, and the steam cock for the coil; a cock for the direct supply of naked steam; a tube for drawing off the liquid matters to settle; and a pipe connecting the cylinder with the reservoir. The cylinder is surrounded on the outside by a stratum of clay, loam, and straw, chopped, to the thickness of 3 inches, so as to prevent the escape of heat. The man-hole is placed at the top of the vessel, a thermometer graduated to 300° Celsius; a barometer; an air-cock; two eye-pieces for observations of the workmen; a pipe 5 inches high leads from the manhole to the supply vessel; this, as well as the hemisphere, has the same coating as the cylinder. The main, or 1st receiving vessel, is a double column, connected internally with the condenser, the outer column receiving the heavy fluid distilling over, which falls back again into the still. The outer column has also the pipe for the reception of fluid destined for the supply vessel alluded to above; a pipe for injecting cold water to condense; also a main cork to disconnect the apparatus, and the air-pump; to this is connected a condensing tube, 100 feet long, and 3 inches wide, cooled by cold water on outside. This tube is connected with air-pumps, the latter having barrels 11 inches wide, and a stroke 13 inches high; these lift water and oil into open casks, where they are separated from each other by repose; the stuffing boxes are made from rings of cast steel. The operation is conducted thus: the tar is deprived of its sulphur by copperas, and then distilled till the liquid is divided in 2 portions—No. 1 and No. 2. No. 1 is oil obtained from the commencement until it reaches 0.870 specific gravity. No. 2 is that obtained from thence on, until the process is completed.

No. 1 is mixed for 4 hours with 6 per cent. of concentrated oil of vitriol; $\frac{1}{8}$ per cent. bichromate potass, and $\frac{1}{8}$ per cent. of muriatic acid.

No. 2 is likewise mixed for 4 hours with 8 per cent. of oil vitriol, $\frac{1}{8}$ per cent. chromate potash, and 1 per cent. of muriatic acid; in two hours the oils are drawn off, and well washed with steam and ley. These washed oils are brought into the reservoir; 1,500 quarts of No. 1 is passed into the apparatus, and the workman then admits steam into the coil; in 20 minutes a temperature of 40° Cent. is attained, when distillation begins—the vacuum is kept at from 25 to 27 inches—violent ebullition or foaming from the presence of water, which only stops at 70° C. The workman looks through the eye-pieces in the vacuum to open the air-pipe, if the fluids should rise too high, and a little skill easily prevents any being carried over. At the beginning of the distillation, cold water is thrown into the condensing fluid by the injecting pipe, to remove any dirt adhering. The first 5 quarts are returned to the reservoir as foul liquid. The heat in two hours is raised to 100° C. Then the gas is lighted, and the apparatus is heated externally by it ; at 120°, the steam is shut off from the coil, and the naked steam cock opened, to maintain a continued motion in the oil ; this pipe is not more than $\frac{1}{4}$ inch wide ; distillation then proceeds quietly, and water is constantly thrown in to keep the pumps clean, and the temperature is raised from 20° to 25° per hour. No. 1 is worked at a temperature of 130° to 140°, and No. 2, 180° to 190°.

Photogen distils over at 200° ; after that the heavy oils are produced, the distillation of which ceases at 250°. The residuum is paraffine, which is removed by a lift-pump into the still. The distilled paraffine is placed in a cellar, and crystallized in moulds.

While the process by the still yields a profit of 92 per cent., that of the vacuum apparatus yields 97 to 98 per cent. ; the 2d distillation reduces the profit of the still to 84 per cent.

Dr. B. Hubner, who has charge of the coal oil manufactory of Messrs. Baumeister & Co., at Bitterfeld, gives the following account of the mode of working, with observations of his own thereon : *

"Brown coal, when distilled in close vessels, commences by breaking up into small pieces, and leaves a coke somewhat resembling gas-coke, though not so dense ; it retains the form of the coal, and is used as fuel for the retorts." The object being to obtain the greatest possible amount of tar, he found it essential that the lowest possible temperature should be exhibited, and that the products formed should be removed from the retorts as quickly as formed : this is attained by using condensing tubes

* Dingler Polytechnisches Journal, Band CXLVI., p. 211. 1857.

not too narrow, and by avoiding as much as possible the use of an hydraulic vessel or main, and by a proper construction of condensers. He describes his process as follows: "I use cast iron elliptical retorts, 8 feet long, 27 inches wide, and 10 inches high; these have this advantage over ⌒ shaped retorts; they are removable when they happen to get burned; the eduction is at the back, the tube for which is at the upper part of the retort, and has this shape (◯) where it leaves the retort, so as to create a large passage, and should be 6¼ inches wide, at a distance of 3¾ inches from the bottom of the retort. The tube has an elbow on it, and has a man-hole in it at the angle, covered with a screw cap.

"Two retorts lie over one fire, with an arched lattice floor between the retorts and the fire: the upper part of the retorts are protected by a layer of ashes, and they (retorts) are so set in the furnace as to be easily put in and taken out."

In Saxony, the distillation is carried on for many days together, by placing several retorts with the fire playing across them, and escaping at the last one; and in its course, it plays on the bottom of one, and on the top of the next, and so on. Low square boxes are the shapes of the retorts, which are filled with coal, so that the coal can be heated both above and below; the lower retort is heated and distilled first. This plan is adopted to save fuel, and to obtain the largest amount of matter worked in the shortest time.

Hubner found his own process to be better than this Saxon one, as regards quantity and quality.

There are many defects in several-day systems; the process of carbonization should be carried on at low temperatures; the lower retorts will always be overheated, while the upper one will not be exhausted : the graduation of the heat is very unequal.

When two retorts only are employed, Hubner recommends that their dimensions be increased above that given by him ; and care should be taken that carbonization goes on all round the retort, from the periphery to the centre. The coal becomes soon agglutinated by the heat, and diminishes considerably in volume ; when the retort becomes heated to redness, the volatile vapors are decomposed, and naphthaline is produced, with a corresponding diminution of paraffine and the lighter oils.

Each of the retorts first-mentioned are filled with 3 bushels (Prussian) of coal, which, when dry, weighs 280 lbs. (Prussian) ; this forms a stratum 3 to 4 inches in depth of the retort, and a free space is thus left for the escape of the vapors, only small portions of which come into contact with the highly-heated portions of the retort, which never quite attain a red-heat ; the period which elapses before perfect carbonization takes place, varies from 8 to 10 hours. Fifty of such retorts at work can use up, in 24 hours, from 360 to 450 bushels, or from 30,000 lbs. to 37,500 lbs. of coal. Slack or fine coal takes longer time to distil than lump coal. It is advantageous to make the slack into lump before using it, because the heat then reaches all portions of the coal more readily through the vacant spaces between the lumps.

In Bitterfeld, the lump coal is separated from the slack by screening, the slack being left for fuel. Hubner conducted a small quantity of low pressure steam through the retorts, so that the steam pipe, finely perforated, being laid at the bottom of the retort, the steam passes through the glowing coal, carrying off the products formed very rapidly, and pure coal is very readily distilled by it. He does not speak of the economy of using steam.

The use of tubes for superheated steam is very expensive, owing to the loss by exposure to heat, and in a new manufacture would not pay, especially in that country, because it is a new manufacture, where simplicity is required in the apparatus used at the outset.

The eduction-tubes enter a common main, 18 inches wide, provided with a man-hole. The main is kept cool in water. Tar and water collect chiefly in this; but little escapes away with the gases, which are passed through a series of condensers, consisting of one pipe placed within another; the gases pass through the outer, which is cooled by water, passing along the inner, also cooled on the inside, and deposits the tar. If the pipes are sufficiently long, wider tubes act most efficiently, for obvious reasons. There is a draught affixed at the point where the gases are drawn off, which are used for heating the furnaces and boiler; the chimney takes the place of the aspirator ○. The use of condensers and purifiers of the gas is objectionable, as increasing the pressure upon the retorts, and preventing the ready escape of the products when formed; a central iron vessel is placed in the centre, and the condensers around it, into which the tar and other products are delivered. The tar and ammoniacal liquor separate from each other in this; by suitable processes, the tar is drawn off clean and free from the water, ready for the still.

The separation of tar from water depends on the relative gravity of the two liquids—in fact, upon the lightness of the tar, and the thickest tar is generally the lightest.

The first light oils come over at 100° Cent., with a

small quantity of tarry water. When the temperature reaches 200° C., there is a momentary cessation of distillation, and a great commotion in the still. When the heat is again pushed, the paraffine oils come over, which readily solidify. Heat is continued until no more fluid product is obtained. When the bottom of the still becomes red, heavy red and pungent vapors arise, along with a yellow fatty tenacious fluid, containing naphthaline, which is the constant companion of products obtained at a high temperature; at this point, a little water is also formed by oxidation of the hydrogen.

The vapors are very injurious to the eyes, and should be conducted off.

It is not economical to distil the tarry liquid by overheated steam.

A still holding 1,000 Prussian quarts, takes 24 hours to distil over.

Tar oils from the Bitterfeld coal, when treated with soda, lose 27 per cent., and the tar oils of the Kœpsner coal, being very hydrogenated, lose 17 per cent.

The crude oils vary in gravity, according to the proportion of creosote. The oils from Bitterfeld range from .890 to .860, while the Kœpsner coal varies from .860 to .840.

Dr. H. Vohl, of Bonn, who has had much experience in the dry distillation of paper coal, recommends a low temperature at commencement, to be raised to a red heat at the end, and that the products be rapidly removed as they are generated. The slate is to be first broken into small pieces of uniform size (not larger than a walnut); if not, they will suffer unequally, the larger pieces not being decomposed when the small ones are fully operated on; they will diminish the profit, and increase the proportion of gas, and produce less oil, because the last portions in the inside of the coal must be decomposed.

Slate, in form of slack, is equally prejudicial, by not allowing the escape of the oily vapors, owing to the close packing of the mass, and thus exposes them to too high a temperature, producing olefiant and marsh gases.

The water contained in slate has an influence on the yield of oil. Vohl obtained from perfectly dry slate, proportionally less light oil than from slates only air-dried, still containing 24 to 25 per cent. of water ; this ratio is that which yields the largest amount of oil. The action of water on slate during distillation is twofold : 1st, it protects the slate from too high a degree of heat ; and 2d, it assists mechanically in carrying off the vapors produced.

Vohl mentions that a loss is produced by too high a heat, causing the paraffine to adhere tenaciously to the gas. The paraffine may be recovered, by passing the gas through a barrel filled with forge-scales, which separates the solid matter ; this is not very remunerative, since the profit from this plan is only 0.1 per cent., and it is not desirable to adopt it, since experience has shown that the danger of explosion arising from condensers is very great, where the method of separating the last portions of oil held by the gas is adopted.

The Rhenish coals sometimes contain poisonous metallic salts ; brilliant crystalline scales of arsenious acid, mixed with sulphide of arsenic and metallic arsenic, form at the elbow of the escape-pipe leading to the main ; and Dr. Vohl states that the slates worked off at the furnaces of Romerickeberge and Stupgen, near Lintz, on the Rhine, owned by A. Wiesmann & Co., contain a large quantity of these poisonous products, and on removing the cap of the retort, a strong smell of arsenic is perceived, and the workmen suffer from colic, ulcers at the root of the nose, of the joints, and an irritable condition of the skin.

The coal oils, as at present sent into the market, are very impure; the demand is so great and disproportioned to the supply, that the manufacturer has neither the necessity nor the time allowed him to redistil or otherwise purify his secondary products arising from distillation of tar. When, however, from a reduced price of animal oil, or any other cause, the demand for oil slackens, then the purification will increase in proportion. In France, where vegetable oils, as rape, camelina, and colza seeds, are extensively grown, the oils of schist, as produced by Selligue and others, are sold in a state of great purity; and in this country, although the public, from motives of economy, may consume coal oils, they will never be used from choice or motives of cleanliness, so long as they are sold in their present condition.

The object of purification is, to separate the viscous, semi-solid, and solid hydro-carbons which are suspended in the lighter oils, and which, from their containing a large percentage of carbon, cannot be made to burn in ordinary lamps without producing smoke, and which produce this annoyance even when present in no large amount in the more volatile liquids.

A redistillation of the oil, carefully conducted, removes much impurity which is retained in the still. The loss of light oil is, however, very large, especially where naked fire is used to heat the still; hence, naked steam, introduced by a coil perforated at the extremity, has been adopted by R. Warrington and others.

Steam, under a higher, but yet moderate pressure, has

also been employed, both alone and in conjunction with external heat, supplied by a steam-jacket, or by fire. The use of steam, in any manner applied (except superheated), is, *cæteris paribus*, a more desirable mode of exhibiting heat than by naked fires. Yet the cost of fittings, boiler, and attendance, may in some situations be such, that the saving effected by steam would be no economy; and in the majority of coal oil factories, the naked fire is applied to the bottom of the iron still.

In addition to re-distillation, the use of chemical agents as purifiers is largely adopted, especially in Europe. Sulphuric acid, caustic soda solution, hydrate of potass and soda, and manganate or permanganate of potass and nitric acid, are the substances most in use: the sulphuric acid, the most powerful, unites with several heavy hydrocarbons, and removes them from the lighter, upon which it has but little action. The manganate of potass and nitric acid, when used, oxidizes several compounds, and thus detaches them from the light oils, and the soda serves the double purpose of neutralizing any acid left in the oils not previously washed out, and also dissolves out the creosote, or carbolic acid.

The purification is effected by chemical means, the impurities not being capable of separation by any means of filtration.

Mansfield, in his patent for obtaining volatile products from tar, describes the purification of benzule by nitric acid, and nitro-muriatic acid. These acids are rarely now employed, sulphuric acid being cheaper and more effective.

The general apparatus for purification does not differ in its essential particulars from that of the purification of gas: a retort or still furnished with refrigerating tubes to conduct away the distilled liquids; hydraulic mains and purifying boxes are the forms of apparatus. In the main, the watery portions separate from the oily and tarry matters, and in the purifying boxes, the less permanent hydro-carbons are broken up and removed.

G. Barry, by his process, patented Sept. 18, 1855, operates in this way. The receiver is placed apart from the retort, and connected by pipes which enter partly into the former: a condenser is provided with refrigerating tubes, condensing the raw oils and ammoniacal waters. The purifiers are made of wooden cases, lined with lead, and provided with agitators. The oils are placed in these after the thick tar has been separated, and treated with 5 per cent. its weight of sulphuric acid. Agitation goes on for 3 hours; the liquid is left to settle for 3 hours, drawn off into a second purifier, placed under the first, when 5 per cent. of their weight of caustic soda, or a sufficient quantity of lime-water is added, and the whole is well stirred for several hours, and then allowed to settle.

After the above process, they are redistilled in the same manner as molasses or rum; after the distillation, the thick liquid tar which remains in the cucurbit, may be converted into a black grease by mixing it with caustic soda; when well stirred, and kept at 75° to 85° F. for two or three hours, saponification sets in, and the matter being run into suitable receivers, forms the paraffinized grease.

The distillation of raw oils is conducted in a cucurbit, placed over the furnace; it has a man-hole for cleansing it, and communicates by a pipe with a coil, from which the products of distillation are discharged into the receiver. The patentee states that the temperature, while distillation is going on, should not exceed from 400° to 600° F.

Hiram Hyde, in his English patent, dated Nov. 24, 1855, describes a method of obtaining volatile oils from petroline, or semi-fluid bitumens, which consists in the rapid application of temperature, beginning about 650°, and passing up to 800°. What is volatilized below 600°, he rejects, as containing too little paraffine; that between the two temperatures is a brown crude oil. This is placed in a leaden vessel, and churned with sulphuric acid for two hours at 90° F. The oil is then drawn off, and agitated with a solution of caustic soda at 30° Baumé, for three hours; a strong solution of manganite or permanganite of potass is then mixed with the oil, agitated for an hour, and left to repose. The oil is then distilled with caustic soda, up to 850°, when the distillate begins to assume a brown hue; the distilled oil is washed with soda solution and jets of steam. By this means, oils having a boiling point of 600° may be obtained. This product is a mixture of hydro-carbons, and is perhaps allied to the coup-oil described as produced by the patented process of Ross.

Schauffele's mode of purifying benzule so as to be unaffected by air or light, remaining always colorless, is, to shake 1 litre of the crude benzule with 100 grammes of ordinary sulphuric acid; allow it to settle for two or three hours; decant the benzule, and shake it anew with another 100 grammes of sulphuric acid; as soon as the separation of the two liquids occurs, the thick colored benzule stratum is decanted off as it floats on the acid, and is shaken with 40 to 50 grammes of dry potash. Sulphate of potash is formed, and the benzule becomes colorless; it is tested, to prove neutrality, and filtered through paper.

In Brooman's English patent for the distillation of

coal-oil, dated Feb. 28, 1856, being a communication from France, retorts and receivers (of common kind) are used for obtaining crude oil, pipes leading from the retorts direct into the receiver. A cucurbit, placed over a furnace, is used for distilling the raw materials. The heat for distillation, it is stated, must not exceed 300° C., (572° F.) The raw oil is distilled by a primary distillation, to get rid of the tar. The oil is brought into contact with 5 per cent. of oil of vitriol, with agitation for three or more hours; then left to settle and draw off into a new purifier; it is then treated with 5 per cent. of caustic soda, or an equivalent of lime-water.

The distilled oil yields a light essential oil (1), whose density at first is 70° of Gay Lussac's Areometer. Distillation is carried on until the liquid has a density of 50°. The first results being light, should be collected separately. With careful distillation, the next batch (2) is collected, until it attains a density of 32° Areometer; this oil may be used for lighting. The heat must be increased for further distillation, when the distilled product (3) will be the lubricating greasy material.

The residue in the cucurbit is a tarry matter. The paraffine may be separated from 2 by cold (10° to 20° C.) which may be obtained by a mixture of ice and sulphate of soda.

Mr. Bancroft, of Liverpool, patented a process for obtaining volatile products from distillation of bitumen or earth-oil, found in Burmah, which consisted in passing high-pressure steam through a still in which the petroleum was placed, the pressure being 50 to 60 lbs. to the square inch; a fire is placed beneath the still until $\frac{1}{4}$th of the original quantity is distilled over, which is eupion nearly pure; this distillate is removed, the fire urged, and steam

AREA OF PRODUCTION. 133

supplied until the remaining 95 parts, or nearly so, have come over, which is eupion combined with hydro-carbons, holding paraffine in solution : at the close, paraffine and pyrole come over largely ; the condensing pipes must be kept at a temperature of 90° F., rising to 120° at the close of the distillation : the residuum in the still contains a large amount of paraffine, which may be obtained by distilling in an iron retort at a low red heat.

Barry, in his patent for decomposing schistose materials, says the heat for the production of oils should never exceed 400° to 600° F.

The area of manufacture of coal-oils is limited, being chiefly confined to the districts where cannel coal can be mined with economy; hence, the States of Kentucky, Virginia, Pennsylvania, Ohio, and Illinois, include at present all the great centres of manufacture. Factories will shortly be established in Missouri, and in every other State where this highly bituminous coal can be obtained. The State of New York is the only exception to the foregoing, the manufacture there being carried on at the seaboard, where the crude mineral (Boghead coal) can be most cheaply delivered. As it is established at the largest market in the U. S., what is overpaid by the use of a costly raw material, is balanced by the reduction of cost of transportation of the refined oil. The following brief and necessarily imperfect notice of the localities of manufacture in this country, contains as complete a list as the author could obtain information about :—

PENNSYLVANIA.—At Darlington Village, Beaver Co., there exists one manufactory of considerable capacity, and three in which the works are on a small scale.

At Darlington Station, or New Galilee, two miles from the village, are the works of the New York Coal Oil Co. This Company rectifies the crude oil. A second manufactory is being raised in this vicinity.

One and a half miles above the mouth of the Kiskiminetas River, on the Alleghany River, in Armstrong Co., the works of Brereton, Williams & Co. are being erected. The revolving retorts of Alter & Hill are introduced. Five retorts are to be set up.

Near Freeport, in Alleghany township, Armstrong Co., on the Alleghany River, are the works of the North American Coal & Oil Co. The works have been in operation since July, 1858. Eight of Alter & Hill's retorts are in operation, 4 large and 4 small. The small ones are 6 feet long and 4 feet in diameter, the large retorts 8 feet long and 6 feet in diameter. Capital invested, $70,000.

The Lucesco Oil Co. commenced operations about the first of April.*

At Rochester is a factory where both the making crude oil and refining are carried on.

At Chester, near Philadelphia, is an establishment for refining the crude oil, being supplied from the western part of the State in which comparatively little in the way of refining is done.

At Pittsburg there is one establishment.

OHIO.—East Palestine, Columbiana Co., has a large factory for crude oil.

At Canfield, Mahoning Co., there are two large establishments, both distilling crude oil, and refining. A third factory is in process of erection.

Close by Steubenville, one medium-sized factory exists, and another is being built.

At Newark are three manufactories of crude oil.

In Cochocton Co., one manufactory is in operation, and six others nearly ready for working.

VIRGINIA.—In Franklin Co., near the Kanawha River, the Union Oil Co., of Maysville, Ky., have their factory for manufacturing crude oil; refining is conducted at Maysville. 800 gallons of crude oil per day is at present produced here, but when all the retorts now being erected are completed, there will be a capability of educing 3,200 gal-

---

* The Lucesco Works, in Westmoreland Co., are probably the largest works at present in operation in the country. The capital invested is $120,000. There are now in working order, ten large revolving retorts placed over as many furnaces, each retort having a capacity of 2½ tons. The mineral is distilled for 24 hours. The crude oil is rectified at the works in stills having a capacity of 2,000 gallons, each armed with agitators, and heated by naked fire; 16 of these stills are erected. The amount of crude oil produced is almost 6,000 gallons per diem.

LOCALITIES OF MANUFACTURE. 135

lons per diem. At the same locality, within five miles of the river, another factory has been started, upon a capital of $30,000.

In the vicinity of Wheeling, some large works are being erected; and on Big Sandy River some crude distillation is carried on on a moderate scale.

KENTUCKY.—The Breckenridge Coal Oil Co. have their extensive works at Cloverport, Ky., where 6,000 gallons per week (May, 1858,) of crude oil are distilled. The coal has already been described; it yields, according to Dr. Peters, for every 100 lbs., 32 lbs. of crude oil.

In Owsley Co., the coal known as "Haddock's Cannel Coal" is extensively manufactured, and yields 55 to 60 gallons of crude oil to the ton.

NEW YORK.—At Brooklyn, on Flushing river, is located the New York Kerosene Oil Co.'s works; both the refining and distilling crude oils are carried out here. The crude oil is distilled from Boghead mineral (coal,) solely in towers or pipes, as patented by Luther Atwood in 1858. Those in operation at present hold 25 tons of coal, and are lighted by anthracite coal, assisted by pine wood at the commencement. The Company are erecting larger retorts than those now in use, being intended to contain 100 tons of coal. The daily produce of crude oil is 1000 gallons.

On the above-mentioned stream, at its mouth, is the factory of the Columbia Coal Oil Co., who heretofore have manufactured crude oil from the Asphalte (or coal) of New Brunswick (the Albert mineral); more lately, however, their attention is almost solely devoted to the refining the crude oils received from the western part of Pennsylvania.

Besides the foregoing, a third establishment is now at work in East Brooklyn.*

* The foregoing list of localities is perhaps imperfect: it is the fullest the author could obtain.

APPENDIX TO CHAPTER II.

DURING the years 1858 and 1859 extensive borings for the purpose of obtaining petroleum or rock oil have been made in Pennsylvania and Ohio. In the former State the most extensive and successful sinkings have been made between the Alleghany river and the western limit of the State; along that river native springs of petroleum have existed which, oozing through the superficial clay, have formed a tenacious, pasty mass. In the vicinity of these springs the artificial wells have been made by sinking a bore deep enough to reach the thin layer of bitumen flowing between the strata. The region now examined may be defined as commencing a short distance above Pittsburg, on the Alleghany river, n Alleghany Co., along the western limit of the State; thence east along the New York State line to the east limit of McKean Co.; thence S. W. to the Alleghany river, where the Conemaugh river joins it. The chief localities are along Mahoning creek, in Armstrong Co.; along the Clarion river, in Clarion Co.; on Oil creek, in Titus, Crawford, and Warren Cos.; at Tidionte, in Warren, near the Alleghany river; along French creek, in Crawford, to Causewago valley.

As the whole of this region is underlaid by what is known to geologists as the coal measures, the petroleum is derived from the natural separation of the bitumen from the carbonaceous portion of the coal, which, oozing upward from faults or fissures in the coal seam, drains off between the strata, and follows the inclination of the latter until it reaches the surface in some denuded portion of the coal bed. This gradual oozing over extensive surfaces yields a large supply of liquid, from which those who sink wells deep enough to reach a thick stratum of petroleum may expect to have an abundant and constant yield, but it is perhaps unnecessary to contradict the popular belief in the existence of a subterranean lake from which these supplies are drawn; such an opinion only could arise from an ignorance of the origin of the petroleum itself. It may be stated that rock oil may be expected to be found in situations where the bituminous coal seams are much disturbed by fractures and dislocations. Where a seam is unbroken no petroleum can escape. The petroleum region, therefore, may be expected wherever coal seams are inclined or tilted at a higher angle than that at which deposition occurred; yet a petroleum spring may not be expected at the eastern extremity of the Pennsylvania coal beds, as they have not only been contorted, but so altered by subterranean heat as to have lost most, and in some parts all, of their bitumen.

The special localities in Pennsylvania where petroleum is sought for, are:

On Oil creek, 2½ miles from its mouth, Messrs. McClintock have a well bored in the fissure of a rock, from which for many years was collected about 15 gallons of oil per diem. The well is 40 feet deep. An engine and pump are being erected. Near this locality Messrs. Crawfords have commenced sinking.

At Titusville, Crawford Co., about 1½ mile below the turn, is the well of Messrs. Drake & Co., the *Pioneer* well. From 10 to 25 bbls. per day are pumped. Bore, 4½ inches in diameter, through 29 feet of earth and 40 feet of rock—total, 69 feet. A surface spring formerly existed here, in which the oil and water rose up through a coarse gravel, and yielded about 12 to 15 gallons per diem. The history of the Pioneer well is as follows:

The Pennsylvania Rock Oil Co. purchased the petroleum spring of Brewer, William & Co., and leased it in 1858 to Mr. E. L. Drake, with the understanding that he should gather the oil at his own expense and pay 12 cents per gallon for it. In May, 1859, Mr. D. commenced boring, and after sinking a shaft 71 feet, a fissure or fault was struck, from which the oil oozed readily.

Within a mile of the town, on bottom land, about a quarter

mile from Oil creek, Messrs. Barnsdale & Co. have sunk a 4 inch bore through 29 feet of earth and 41 feet of rock—total, 70 feet: a considerable supply is promised here.

Within 30 rods of the preceding is the well of Messrs. Williams & Co. The boring was conducted in clay for the first 96 feet, when rock was reached: a 5 inch cast-iron tube was sunk. The following statement gives the total depth and character of the rock bored:

	Distance
One foot muck,	1
Five feet blue clay,	6
Forty-three feet mixed gravel,	49
One foot blue clay and sand,	50
Six feet sand, clay, and shales,	56
Twenty-six feet fire clay, striking nodules at bottom,	82
Four feet sand, gravel, and pebbles,	86
Nine feet gravel and fine sand,	95
One foot fine gray sandstone,	96
Three feet of shale rock, striking seam of gas,	99
One foot soap rock, with water and oil,	100
Four feet soap rock, with oil more and more plenty,	104
Eleven feet soft blue shale, with additional supply of oil,	115

One mile below McClintock's, Messrs. Ewing & Shugert are boring with an engine—have reached 30 feet. In this neighborhood W. Stewart & Co. are also sinking.

Eleven miles below Titusville, Messrs. Kellogg & Co. have sunk a well 90 feet, with a $4\frac{1}{4}$ inch bore. Two barrels per day are obtained. They propose to sink deeper.

In this vicinity, Aleen, Chase & Co. and Brown, Mithel & Co. are sinking.

Two miles above Titusville is the well of the Kerrs. In the vicinity of the town, Moore, Chase & Co. have sunk 130 feet, and reached a rich layer of oil. Three-fourths of a mile below Moore's is the well of Mead, Rouse & Co., 96 feet deep, and close by, that of Williams, Tanner & Co., 110 feet deep. Brine is pumped up in the last well.

On the opposite side of the creek are the wells of Donaly,

Kier & Co., and of Allen & Johnston. The latter have found oil at 130 feet.*

One-fourth mile below the Pioneer well is that of Crossley, Sloan & Co., 108 feet deep; and on the hill opposite, Ullman & Co. sunk a considerable depth, but were prevented proceeding by the leakage of gas.

One mile below Crossley's, Fletcher, Stockspole & Co. have reached an abundant oil supply at 90 feet. Around this vicinity are many borings as yet uncompleted.

At Tidionte, Warren Co., Messrs. Dennis & Co. are boring, about 1½ miles from the town and the Alleghany river, on Gordon's run; bore 2½ inch, and 63 feet down in rock, which is within 3 feet of the surface. About 1 gallon per day is collected. At the mouth of this run, Messrs. King & Co. have sunk a well 6 feet diameter to 17 feet deep, where rock is reached. The oil collected, about 3 gallons per day.

Near Tidionte, the Lennine Exploring Co. are sinking five wells, 8 inches in diameter, and from 17 to 63 feet deep. Traces of oil are found in two of them.

At Tarentum, Alleghany Co., are three borings, made originally for brine, and still yielding salt water. The oil comes up with the brine, and separates completely by subsidence, and communicates no flavor to the salt. The borings are 450 feet deep, the brine coming from the lowest point, and the oil from about 350 feet, or 100 feet above the brine spring. Of the three wells, that of Peterson & Co. yields about 10 bbls. in 24 hours; Kier's about 3 bbls.; and Peterson Sen.'s, about 1 bbl. per 24 hours. The first-named well yielded brine for 20 years without a trace of oil, when the diameter was increased from 3 inches to 4; it then began to yield oil in the amount of 3 bbls. for 24 hours. From time to time the diameter of the bore was increased, the supply of oil increasing until the diameter reached seven inches, its present size, with the yield above given. Many of the old salt wells about Tarentum are now being deepened with the hope of obtaining oil from them.

At Tarentum, L. Peterson & Co. are sinking a shaft 4½ feet

---

* For some of the information received, we are indebted to the *Commercial Gazette*, of Titusville.

by 8, which they propose to reach 400 feet in depth, so as to cut through on a large scale the oil-bearing stratum. Not more than 50 feet is at present sunk.

At Franklin, Venango Co., the Franklin Co. have bored 40 feet through rock, having commenced at the bottom of an abandoned well. The total depth is 60 feet, with a 6 inch bore; this is situate about 20 rods from French creek. Two miles below Franklin, Stewart & Co. have reached 90 feet with a 4 inch bore, and obtained oil. Two miles above Franklin, on the Alleghany river, Messrs. Fulton & Co. are sinking. At the mouth of Oil creek, Messrs. Arnold & Co. have sunk 325 feet, and obtained oil. About a mile from this, on the east bank of Alleghany river, a company have sunk 60 feet in rock, but have not yet reached oil.

Oil has been discovered in the vicinity of the mouth of Deer creek, on the Clarion river, on the Packer property, now in possession of Mr. Whitehill. Oil has been found on the Clarion, between the old bridge and Russell's mill, and near Shippenville springs have also been discovered, rendering the excitement intense. The McCormick well yields about a gallon of oil each minute. The sum of $400,000 has been offered for the property.

SYNOPTICAL RESUME

OF

PATENTED IMPROVEMENTS HAVING REFERENCE TO THE DISTILLATION OF OILS FROM COALS, BITUMENS, AND SCHISTS.

I. AMERICAN PATENTS.

1852. *March* 23.—JAS. YOUNG. Improvement in making Paraffine Oil; (English patent dated Oct. 7th, 1850;) claims "obtaining paraffine oil, or an oil containing paraffine, and paraffine from bituminous coals, by treating them in the manner heretofore described;" distils the coal at a low red heat; treats distillate with sulphuric acid, and soda solution, redistils, and repurifies, and distils a third time.

1853. *March* 29.—LUTHER ATWOOD. Process of preparing Paranaphthaline Oil from the distillate of Coal Tar; collecting the products at certain fixed temperatures; calls the product "Coup-Oil."

1853. *September* 27.—WM. BROWN. Preparing Paraffine Oil, Lubricating Oil, and Eupion, from Coal or other bituminous matter; claims the use of super-heated steam, as specified, for separating the products; also claims the modes of separating Eupion, Paraffine, and Lubricating Oil from each other.

1854. *June* 27.—ABM. GESNER. Production of Kerosene Oils from Maltha and other bituminous substances, by subjecting them to dry distillation, at a heat not exceeding 800° F. The liquid distillate divides into 3 strata. The upper stratum is drawn off and redistilled; this 2d distilled is purified and distilled to produce Kerosene A: analogous liquids, obtained by similar treatment with varying temperatures, yield Kerosene B and C. Claims the liquid Kerosene.

1855. *March* 27.—ABM. GESNER. Improvement in processes for making Kerosene, by dry distillation, at the lowest temperature at which Kerosene will volatilize. The fluid is obtained by processes similar to those described in the foregoing description; claims obtaining Kerosene from bituminous substances, by subjecting any of them to dry distillation, rectifying the distillate by treating it with acid and freshly-calcined lime, and then submitting it to re-distillation. as set forth.

1856. *August* 12.—L. & W. ATWOOD. Improvement in production of oil from Cannel Coal, so as to form a lubricating oil, consisting of Paraffine dissolved in Eupion, or light oils obtained in the first distillation. This oil boils at 600° F., is fluid at 32° F., and of a density of .864 at 60°; claims the oil produced having the properties set forth.

1856. *August* 12.—L. & W. ATWOOD. Trinidad Pitch, or Barbadoes Tar is distilled, and the product is again distilled: this distillate is purified by sulphuric acid, and afterward caustic soda, and finally by permanganite of potass, or soda; the fluid is then finally distilled. This fluid boils at 600° F., is fluid at 32° F., and has a density of .900. Claims the manufacture and use of the oil described.

1856. *September* 2.—CUMMINGS CHERRY. Improvement in apparatus for purifying oil obtained from Mineral coal. The crude oil is distilled in a horizontal retort furnished with copper heads and receiver, into which the distillate rises, whence it is driven into the rectifying chamber, furnished with trays, on which is placed a stratum of unslacked lime; the vapors are then passed into a condenser and cistern, in which muriatic acid diluted is made to act on the liquid by means of agitation; after repose and decantation, the fluid is subjected to milk of lime. The oil is then drawn off and pumped into a boiler, where it is exposed to the direct action of steam. Claims the arrangement of the retort, combined with the copper heads, the rectifying chamber, the steam conduits, and the agitating apparatus.

1856. *September* 2.—CUMMINGS CHERRY. Improvement in apparatus for distilling crude oil from Mineral coal. The coal is fed into an upright retort, having a closed top, and open at the lower extremity, surrounded on inside with fire-tiles; the bottom of the retort is immersed in water. An agitator, or stirring rod, with small lateral projections attached, is fixed vertically in the retort, to keep the materials at a uniform temperature. Claims providing upright retorts with a closed top, and opening at the bottom, to be immersed in water, as set forth.

1856. *September* 2.—CUMMINGS CHERRY. Improvement in the preparation of drying oil from oils extracted from bituminous minerals. The purified oil is boiled with litharge and common resin. Claims preparing the oil as set forth.

1856. *December* 16.—RICHARD SCHRODER. Improvement in apparatus for Coal-Oil. The coal is distilled in small upright retorts of fire-clay, closed at top, and set in a furnace so as to be surrounded with flame and fire, with pipes leading from it at different heights, so that the oils may be separated from each other while distilling, and not require subsequent rectification. Claims, constructing the retort or generator with openings of different heights, as shown, for the purpose of obtaining oil of different qualities, as set forth.

1858. *April* 27.—DAVID ALTER and S. A. HILL. Re-issued *February* 8, 1859. Improvement in retorts for obtaining volatile liquids by dry distillation of Shale, &c.; distils the coal, &c., in a cylindrical retort of cast iron or other metal, which rotates on an axle prolonged at each end; to the front extremity is attached the wheelwork needed to produce revolution; the axle at the back of the retort is hollow, allowing the liquids and gas to escape into the condenser.

1858. *June* 15.—T. D. SARGENT. Improvement in Revolving Retorts for distillation of volatile oils from coal—a clay retort, placed horizontally, and worked so as to revolve to a limited extent; that is, when moved round two-thirds of its periphery, it returns back. Claims, a retort made of clay, and having a revolving motion when in action.

1858. *August* 10.—T. & W. B. McCUE. Improvement in apparatus for extracting oil from coal. Uses a revolving horizontal cylindrical retort, which passes ¼ of its periphery, and then returns; the retort is furnished with elevated plates or ribs running parallel to the long axis of the retort; these aid in preventing the coal accumulating in a mass at the lowest part of the retort. Claims the elevated plates or ribs described.

1858. *October* 19.—LUTHER ATWOOD. Improvement in processes for obtaining volatile oils from coal, wood, &c. The coal, &c., is distilled in a tubular or cylindrical vertical retort, or tower, open at the upper extremity, by which the retort is fed; the eduction pipe is placed at the lower part of the tower, and leads to the condenser or tank; from this latter, a curved pipe leads to the worm; between the tank and the worm a steam jet nozzle is affixed, so that aspiration may be effected, by which the current of distilled products is directed downwards: a fire is first kindled at the open mouth of the retort when filled; the aspirator is then put in action, when the distillation downwards goes on slowly without interruption.

1858. *December* 28.—LUTHER ATWOOD. Improvement in apparatus for distilling coal. The process is that above described. Claims the combination and arrangement of a distilling tower and receiving vessel, as described, with a steam blast, or its equivalent, for producing an increased current, as set forth.

1858. *December* 28.—LUTHER ATWOOD. Improvement in manufacture of Pyrogenic Oils; places the substances to be distilled on the sole of a reverbatory furnace of a peculiar construction, so that the sole may be heated as well as the arch. Claims forming Oleaginous Vapors from substances yielding pyrogenic oils, by the action of the heat of a properly regulated current of the products of combustion passing over and above the surface of the mass operated on, with or without the aid of external heat, as described.

1858. *December* 28.—LUTHER ATWOOD. Apparatus for decomposing wood, bones, &c. This apparatus is adapted for dry distillation in general, and in principle is the same with that patented by the applicant, October 19. The fire, in this case, is external to the tower, and the flame, &c., is conveyed to it through flues; a steam aspirator is used here also. Claims the combination of the distilling tower with the fire-place, when so arranged as to supply products of combustion by a downward draught through the fire-place, as set forth.

1859. *January* 11.—JAMES O'HARA. Improvement in apparatus for distilling oils from coal; a vertical retort is used, having a feed-pipe and eduction pipe at the upper end; the retort is placed in a fire-place, and supported on flanges attached near the upper part of the side. An Archimedean screw is placed in the centre of the retort, for stirring the coal—there is room left between the plates of the screw and the inner wall of the retort for the coal to drop down. Claims, in an upright retort, the use of a revolving screw of less diameter than the inside of the retort, so as to allow of the ascent as well as the descent of the coal at the sides of the retort.

1859. *January* 25.—E. N. HORNER. Improvement in processes for extracting oils from coal. Claims the use of a compound of cream tartar, salt, and lime placed in the bottom of the condenser, to separate the steam from the oil, to condense the vapors, and to eliminate sulphurous acid gas.

1859. *February* 1.—N. B. HATCH. Improvement in retorts for distilling oils from coal. The coal is fed into a semi-globular-shaped flat-bottomed still, or retort, through a hopper, and while being distilled, is kept in motion by a sweep-bar, or vertical arm, with horizontal shafts attached, which are furnished with metallic plates, so as to sweep the bottom of the vessel while in motion; eduction pipes are placed at the lower margin of the vessel on a level with its bottom. Claims the application of a sweep-bar, or arm, with plates attached, operating so as to push or spread the material over the floor, and at intervals discharge some continuously by openings at or near the edge of the retort, as set forth.

1859. *February* 15.—JOHN NICHOLSON. Re-issued *May* 3. Improvement in retorts for distilling coal-oil; a cylindrical-horizontal retort is fitted with a shaft travelling through the long axis, furnished with agitators or arms having curved blades. At the extremities of the retort, openings exist: 4 at one end for the attachment of supply and discharge pipe, and at the other end, 4 exit pipes. Claims the shaft or agitator armed with curved blades; also the openings at the ends of the retort, as described. On a re-issue, a claim to the use of straight blades also, was secured.

1859. *February* 22.—LUTHER ATWOOD. Improvement in apparatus for destructive distillation. This form of apparatus is but a variation of that already described; the fire is external to the tower, and the heated air enters the upper part of the tower by a bent flue; the combustion is carried on by aspiration. Claims the arrangement and combination of the combustion tower, the distilling tower, and the steam blast or its equivalent, as set forth.

1859. *March* 29.—JAS. GILLESPIE. Improvement in coal-oil retorts. Uses a revolving horizontal cylindrical retort, with shaft passing through its long axis; the eduction pipe is formed by the hollow extremity of the axle; in order to keep the mouth of the eduction pipe always in an upright position, it is secured by pins surrounding the journal. Claims, securing the hopper-cup with pins, or their equivalents, surrounding the journal, with the square-headed shaft.

1859. *March* 29.—LUTHER ATWOOD. Improvement in apparatus for destructive distillation. Combines a vertical distilling tower, as before patented, having an air-tight cover and feed-opening, with a condenser and adjustable draft passage, furnished with a sliding door or damper, so as to regulate the passage of air to the fire; distillation goes on by an upward current.

1859. *April* 19.—WILLIAM SMITH. Improvement in coal-oil retorts. Uses a horizontal cylinder retort, furnished with a hollow shaft having hollow arms attached, so that a current of air or water may be driven through to cool the retort.

1859. *May* 31.—JOSEPH E. HOLMES. Improvement in coal-oil retorts. Uses an L shaped retort, with a central perforated pipe attached to the cover, and suspended from it, to allow of the escape of the vapors, leaving an open space beneath it, through which the material may be removed. Claims the perforated pipe, as set forth.

1859. *May* 31.—R. HAZLETT and T. H. HOBBS. Improvement in coal-oil retorts; retort also useful for general purposes of distillation. The base of the retort is flat, or rectangular, and the sides convex. The fire is applied to the lower portion only. The retort has a false bottom, or charger, for holding the coal; an air-chamber or space is allowed between the pan and the bottom of the retort, that the coals may not be burned. The retort has conduits or gutters running along the lower part of the sides of the retort. Claims the shape of the retort, and the charger.

1859. *March* 29.—Jos. E. HOLMES. Protects the hollow journal eduction pipe from entrance of coal, &c., by fitting on an elbow inside the retort, and carrying it to the upper portion of the retort; also adapts a perforated steam pipe passing through the journal, and diffusing steam through the coal.

1859. *May* 31.—WM. G. W. JAEGER. Improvement in retorts for distilling coal-oils. A cucurbit shaped retort, with a flat bottom, having side-channels and trap-openings, or water-valves, by which the heavy oils or tar may be removed, while the lighter oils pass off by the neck. Claims the side-channels and trap-openings, also the try-hole in combination with the retort described.

1859. *June* 21.—H. P. GINGEMBRE. Improvement in apparatus for destructive distillation. Uses an L shaped retort, combined with charging-boxes, crusher, and discharging-tube, as described, capable of being subjected to a degree of temperature higher at the horizontal than at the vertical end; atmospheric air being excluded. The crusher is placed in the retort between the point of greatest and least heat.

1859. *June* 28.—W. G. W. JAEGER. An improvement in the mode of condensing vapors of oil, by introducing between the retort and condensing worm a large surface condenser of special construction; attached to this is a fan-blower, an escape-pipe, and a trap-opening. Claims the novelty in the apparatus.

1859. *June* 30.—JOHN L. STEWART. Improvement in retorts; uses a revolving web-retort, with induction and escape-pipes at one end; a coal-feeding endless apron, which carries the coal twice through the retort. A water-trough and endless carrier to remove the coke, is attached

1859. *August* 2.—WM. T. BARNES. Improvement consists in attaching to a coal oil apparatus an automaton dust-clearer, consisting of a series of levers and rod, operated by a cam or otherwise. Spiral or screw flanges are adapted to the head of the retort for pushing the material away from the hole in the journal.

1859. *August 2.*—HENRY PEMBERTON. In the refining of coal-oil, claims recovering the sulphuric acid used in the process, by treating the residuum with hot water, steam, or otherwise.

1859. *August 2.*—WM. T. BARNES. Claims a tube provided with tubular arms, made to revolve, and connected with a water supply, as set forth. Claims also the arrangement of the water tanks.

1859. *August 16.*—H. P. GINGEMBRE. Claims the continual progression and gradual destructive distillation of coal or other bitumens.

1859. *September 20.*—MORRIS L. KEEN. Claims, in the distillation of coal-tar, the employment of additional heat, at or near the surface of the coal-tar or other similar hydrocarbon, when used in combination with pressure in the boiler to prevent the tarry foam rising up in the vessel.

1859. *November 29.*—MATTHEW HODGKINSON. Claims a stationary retort, armed with knives whose edges are at right angles with the shaft passing through it, by which the coal is broken and powdered more economically.

1860. *January 3.* F. W. WILLARD. Furnishing coal-oil retorts with internal false or extra heads at either end of the retort, and held at proper distances by means of stays or studs, as set forth; the intervening space being filled with clay or other non-conducting material.

II. EUROPEAN PATENTS.

Under this head it has not been deemed necessary to give an abstract of each patent, as the descriptions are extensive, and the claims numerous, in the great majority of the patents; they are classified here under the several general natures of the inventions claimed.

ENGLISH.*

a. General Manufacture.

1746. Aug. 7. H. Haskins.
1781. April 30. Earl of Dundonald.
1833. Jan. 29. Richard Butler.
1842. Mar. 4. T. A. W. Count de Hompesch.
1850. Oct. 7. Jas. Young.
1851. Nov. 22. Jas. Gilbee.
1852. Nov. 5. Earl of Dundonald.
1852. Nov. 5. G. Shand, and A. McLean.
1853. Jan. 13. Wm. Brown.
Feb. 4. Jno. Perkins.
March 18. Geo. Rose.
March 31. W. A. P. Aymard.
April 22. C. M. T. du Motay.
July 5. John Fall.
July 25. Warren de la Rue.
August 18. Jno. Perkins.
August 23. Wm. Brown.
Oct. 12. E. J. Maumene.
Dec. 9. J. Chisholm.
Dec. 27. F. C. Calvert.
1854. Feb. 4. Wm. Little.
Feb. 15. G. F. Wilson.
March 3. Wm. Brown.
May 10. Rees Reece.
June 23. D. C. Knab.
July 26. P. A. Godefroy.
Dec. 23. Warren de la Rue.
1855. Jan. 22. Wm. Kilgour.
Feb. 7. Edward Davies.
Sept. 4. W. de la Rue.
Nov. 24. H. Hyde.
Dec. 5. Davies, Syers & Humphrey.
1856. Jan. 3. Herman Brambach.†
April 10. P. Bancroft, and S. White.
April 22. A. E. Beach.
May 15. J. G. Simpson and W. Thompson.
Sept. 10. Stephen White.
Dec. 6. James Perry.
1857. Jan. 8. T. W. Keats.†
Jan. 12. G. F. Wilson.
Jan. 28. G. F. Wilson.
1858. Feb. 24. F. Puhla.
Feb. 24. F. Puhls.
April 6. W. Ziernozel.
May 26. J. Stuart.

b. Apparatus for Distillation.

1852. Dec. 28. Edward Mucklow.
1853. Jan. 24. D. C. Knab.
August 13. A. M. M. de Bergerin.

1858. August 22. A. E. L. Bellford.
Oct. 21. J. F. F. Challeton.
Nov. 2. F. A. Gatty.
Dec. 5. Edward Lavender.
Dec. 20. Paul Wagenmann.
1854. Jan. 6. II. H. Edwards.
July 14. A. P. Price.
Sept. 14. G. F. & F. J. Evans.
Dec. 15. F. Archer & W. Papineau.
1855. May 29. E. J. Lafond and A. de Chateau Sillard.
July 18. John Ellis.
Sept. 18. P. G. Barry.
1856. Feb. 28. P. G. Barry.
Sept. 10. Stephen White.
Dec. 6. W. H. Bowers.
1857. March 31. T. E. D'Arcet.
August 5. Sebastian Bottiere.
Sept. 9. Edward Lavender.
Oct. 22. A. H. C. Chiandl.

FRENCH.‡

a. General Manufacture.

1848. Nov. 17. Lacarriere.
1850. April 23. Lacarriere.
July 29. Ferrand.
1851. Feb. 15. Bourdeux.
1852. April 17. Bourdeux.
1853. Jan. 15. Potsat, Knab and Mallet.
May 25. L'Isle de Sales.
Sept. 7. Chatelan and Encontre.
Oct. 8. Challeton.
1854. Jan. 12. L'Isle de Sales.
June 27. Burdin.
1855. Dec. 24. Renaud.
1856. April 24. Beach.
Nov. 26. Tripon.
1857. Jan. 10. Camus and Messillier.

b. Apparatus.

1850. Jan. 16. Brehot.
Jan. 29. Maillart.
March 25. Maillart and Ganneron.
April 29. Lahore.
1851. Nov. 12. Girandel.
1853. Feb. 6. Malo, Prosper and Hugues.
April 4. Buran.
June 20. Humbert.
1854. April 13. Lacasse and Millochau.
July 3. Sauvage.
Sept. 15. Challeton.

---

* For descriptions, consult "English Specifications of Patents, by Bennet Woodcroft," published by Royal authority, London.
† Thus marked are void specifications, not being completed.
‡ Patents in force not published by the Government. Those which have expired may be consulted in the Catalogue des Brevets d'Invention, published by the French Government.

INDEX.

Agitators in retorts, 104.
Albert mineral, 16.
Albert mineral, oils from, 84.
Alter & Hill's process, 113.
Alloys as heating agents, 101.
Alloys, table of fusibility of, 101.
American Patents, list of, 136–140.
American coals, 30.
Ampeline, 62.
Ammonia from heat, 89.
Aniline, 65.
Anthracene, 70.
Archimedean Screw in retort, 105.
Asphalt from turf, 88.
Aspirators, use of, 106.
Atwood's mode of distilling, 111.
Baths, metallic, 101.
Benzule, 42, 58, 59.
Bitumens, 31.
Bitumen, analysis of, 31.
Bitumen in coal, 21, 22.
Bitumen, nature of, 33, 34, 35.
Bitumen, varieties of, 31, 32.
Bitterfeld, distillation at, 124.
Boghead coal, 25, 26.
Boghead mineral, 25.
Bonn, manufacture at, 116–120.
Breckenridge coal, 24, 25.
Brown coal, distillation of, 122.
Cannel coal, localities of, 28.
Carbolic acid, 63.
Cement for clay retorts, 96.
Chervau, C. & P., notice of patent of, 13.
Chimney, distillation upward, 109.
Chimney, distillation downward, 110.
Chrysene and Pyrene, 70.
Clayton, Dr., experiments, 6, 7.
Coal, analyses of, 33.
Coal, chemical change in, 18, 19.
Coal, definition of, 17.
Coal, slow decomposition of, 72.

Coal, influence of pressure, 73.
Coal, microscopic examination, 17.
Coal, varieties of, 23.
Coal, distillation of, 53
Coals, nature of, 15.
Coals, fat, 73.
Coke, 36, 40.
Coup oil, 65, 66.
Crane furnace, 98.
Creosote, 63.
Crude oils, purification of, 121, 130–132.
Cumene, 61.
Destructive distillation, 35, 42, 46.
Distillation in towers, 106.
Distillation, substances formed by, 36.
Dorsetshire shale, 82.
Eaton, Professor A., remarks by, 114.
English patents, list of, 141.
Escape pipes, 100.
Eupion, 90.
French patents, list of, 141.
Furnace for peat, 98.
Gases of combustion, heat of, 109.
Germany, manufacture in, 114, 115.
Growth of the art, 16.
Hales, Dr., remarks, 8.
Hamburg, factory at, 114.
Hatcheltine, 31, 34.
Heat, application of, 92, 93, 94.
Hubner, on distillation of coal, 122-125.
Hydrocarbons, table of, 34.
Hydrocarbons isomeric with paraffine, 69
Hydrocarbons, fossil, 74.
Hydrogen, carbide of, 39, 42.
Irish Peat Co. works, 88.
Lewitte, notice of patent of, 12.
Lignite, 29.
Lignite, distillation of, 54, 126, 127.
Light oils, 60,
Mansfield, notice of patent of, 9.
Mansfield purification of benzule, 120.

Manufacture, area of, 133.
Manufacture, extent of, 134, 135.
Manufacture, localities of, 134, 135.
Middletonite, 31, 34.
Naphtha, 31, 58.
Naphtha, density of, 32.
Naphtha in Boghead coal, 27.
Naphtha from schists, 82.
Naphthalin, 70.
Naphthalin, formation of, 93.
Newberry, Dr., views on cannel coal, 27.
Northern, Mr., experiments, 8.
Oils, produce of, 55,
Oils from bituminous schists, 82.
Oils, purification of, 118, 121, 128, 129.
Oils from turf, 86, 87.
Oils of wood tar, 90.
Organic substances, decomposition of, 36.
Ovens, brick, 97.
Ovens, carbonizing, 97.
Ozokerite, 31, 34.
Paraffine, production of, 67.
Paraffine, properties of, 68.
Paraffine, purification of, 117.
Paraffine, when formed, 93.
Paraffine, recovery of, 118, 127.
Paraffine, fossil, 31.
Paraffine of turf, 88.
Peat, origin of, 32.
Peat, products of distillation, 89.
Peat produce in oils, 89.
Petroline, 82.
Photogen, of Wagenmann, 118, 122.
Photogen from turf, 87.
Photogenic oils, 62.
Picamar, 91.
Pipes, distillation in, 110, 111.
Pittacal, 91.
Pouillet, table of temperatures, 94
Pyrene, 71.

Pyroxanthin, 91.
Reichenbach, notice of experiments of, 11.
Resins, fossil, 74.
Resins, formation of, 75.
Retorts, form of, 95–100.
Retorts, shape of, 100.
Retorts, position of, 99, 100.
Retorts, vertical, 99.
Revolving retorts, 102.
Revolving retorts, value of, 103, 104.
Rhenish coals, 127.
Saxony, distillation in, 123.
Schists, distillation of, 82.
Selligue, M., process of distilling, 82.
Shale, bituminous, products from, 82, 83.
Slate, posidonian, oil of, 83.
Steam, effect on distillation, 40.
Steam open, in distillation, 107.
Steam, superheated use of, 108.
Steam, purification by, 129.
Tar, amount of, 48.
Tar, nature of, 48.
Tar, constitution of, 56, 57.
Tar, production of, 49.
Tar, constituents of, 58.
Tar from Cannel coal, 51.
Tar from turf, 86.
Temperature for distilling, 39–94.
Temperatures, table of, 94.
Toluene, 60.
Towers, distillation by, 106, 110, 111, 112.
Turf, mode of growth, 32.
Turf, distillation of, 52, 85, 86.
Volatile oils, distillation of, 38.
Vohl on Lignite, 126, 127.
Wagenmann, process of, 115, 116.
Wagenmann's mode of distilling, 117, 120.
Wood, carbonization of, 40.
Young, James, notice of patent of, 10.

www.ingramcontent.com/pod-product-compliance
Lightning Source LLC
Chambersburg PA
CBHW022125160426
43197CB00009B/1158